ANIMALS

From Mythology to Zoology

DISCOVERING
the
EARTH

ANIMALS

From Mythology to Zoology

Michael Allaby

Illustrations by Richard Garratt

Facts On File
An imprint of Infobase Publishing

ANIMALS: From Mythology to Zoology

Facts On File, Inc.
An imprint of Infobase Publishing
132 West 31st Street
New York NY 10001

Library of Congress Cataloging-in-Publication Data

Allaby, Michael.
    Animals: from mythology to zoology / Michael Allaby; illustrations by Richard Garratt.
        p. cm.—(Discovering the Earth)
    Includes bibliographical references and index.
    ISBN-13: 978-0-8160-6101-3
    ISBN-10: 0-8160-6101-7
    1. Animals—Popular works. 2. Animals—Mythology—Popular works. 3. Zoology—Popular works.
I. Title.
    QL50.A395 2010
    590—dc22   2009006194

Facts On File books are available at special discounts when purchased in bulk quantities for businesses, associations, institutions, or sales promotions. Please call our Special Sales Department in New York at (212) 967-8800 or (800) 322-8755.

You can find Facts On File on the World Wide Web at http://www.factsonfile.com

Excerpts included herewith have been reprinted by permission of the copyright holders; the author has made every effort to contact copyright holders. The publishers will be glad to rectify, in future editions, any errors or omissions brought to their notice.

Text design by Annie O'Donnell
Illustrations by Richard Garratt
Photo research by Tobi Zausner, Ph.D.
Composition by Hermitage Publishing Services
Cover printed by Creative Printing
Book printed and bound by Creative Printing
Date printed: December, 2009
Printed in China

10 9 8 7 6 5 4 3 2 1

This book is printed on acid-free paper.

CONTENTS

PREFACE

Almost every day there are new stories about threats to the natural environment or actual damage to it, or about measures that have been taken to protect it. The news is not always bad. Areas of land are set aside for wildlife. New forests are planted. Steps are taken to reduce the pollution of air and water.

Behind all of these news stories are the scientists working to understand more about the natural world and through that understanding to protect it from avoidable harm. The scientists include botanists, zoologists, ecologists, geologists, volcanologists, seismologists, geomorphologists, meteorologists, climatologists, oceanographers, and many more. In their different ways all of them are environmental scientists.

The work of environmental scientists informs policy as well as providing news stories. There are bodies of local, national, and international legislation aimed at protecting the environment and agencies charged with developing and implementing that legislation. Environmental laws and regulations cover every activity that might affect the environment. Consequently every company and every citizen needs to be aware of those rules that affect them.

There are very many books about the environment, environmental protection, and environmental science. Discovering the Earth is different—it is a multivolume set for high school students that tells the stories of how scientists arrived at their present level of understanding. In doing so, this set provides a background, a historical context, to the news reports. Inevitably the stories that the books tell are incomplete. It would be impossible to trace all of the events in the history of each branch of the environmental sciences and recount the lives of all the individual scientists who contributed to them. Instead the books provide a series of snapshots in the form of brief accounts of particular discoveries and of the people who made them. These stories explain the problem that had to be solved, the way it was approached, and, in some cases, the dead ends into which scientists were drawn.

There are seven books in the set that deal with the following topics:

- Earth sciences,
- atmosphere,
- oceans,
- ecology,
- animals,
- plants, and
- exploration.

These topics will be of interest to students of environmental studies, ecology, biology, geography, and geology. Students of the humanities may also enjoy them for the light they shed on the way the scientific aspect of Western culture has developed. The language is not technical, and the text demands no mathematical knowledge. Sidebars are used where necessary to explain a particular concept without interrupting the story. The books are suitable for all high school ages and above, and for people of all ages, students or not, who are interested in how scientists acquired their knowledge of the world about us—how they discovered the Earth.

Research scientists explore the unknown, so their work is like a voyage of discovery, an adventure with an uncertain outcome. The curiosity that drives scientists, the yearning for answers, for explanations of the world about us, is part of what we are. It is what makes us human.

This set will enrich the studies of the high school students for whom the books have been written. The Discovering the Earth series will help science students understand where and when ideas originate in ways that will add depth to their work, and for humanities students it will illuminate certain corners of history and culture they might otherwise overlook. These are worthy objectives, and the books have yet another: They aim to tell entertaining stories about real people and events.

—Michael Allaby
www.michaelallaby.com

ACKNOWLEDGMENTS

All of the diagrams and maps in the Discovering the Earth books were drawn by my colleague and friend Richard Garratt. As always, Richard has transformed my very rough sketches into finished artwork of the highest quality, and I am very grateful to him.

When I first planned these books, I prepared for each of them a "shopping list" of photographs I thought would illustrate them. Those lists were passed to another colleague and friend Tobi Zausner who found exactly the pictures I felt the books needed. Her hard work, enthusiasm, and understanding of what I was trying to do have enlivened and greatly improved all of the books. Again I am deeply grateful.

Finally, I wish to thank my friends at Facts On File, who have read my text carefully and helped me improve it. I am especially grateful for the patience, good humor, and encouragement of my editor, Frank K. Darmstadt, who unfailingly conceals his exasperation when I am late, laughs at my jokes, and barely flinches when I announce I am off on vacation. At the very start, Frank agreed this set of books would be useful. Without him they would not exist at all.

INTRODUCTION

People share the world with a bewildering variety of nonhuman animals, and throughout history people have been trying to understand the animals around them. Clearly, some of them resemble humans in certain respects, and it seems entirely natural to attribute to them feelings, thoughts, and motives that a human might have. Some animals help humans, some steal from them, some are dangerous, and all of them are interesting. Where did animals come from? Why do they exist? This book describes a few of the changing ideas that have shaped the way people see the world of animals, and it tells of some of the research that has revealed their physiology and explained their evolution and behavior.

Animals, one volume in the Discovering the Earth set, begins the story in ancient Greece, at a time when thinkers were trying to construct a coherent image of the universe. They needed to make sense of the world around them, and the earliest theories explaining the origins of animals were drawn from the experiences of hunters. As they debated and argued, speculated and theorized, the Greek philosophers came to realize that careful observation of what really happens is the only basis for a true understanding of the natural world. With that realization they laid the foundation of what would become a scientific approach to phenomena.

Since ancient times, animals have also been symbols of human qualities. Lions are brave, foxes sly, serpents treacherous. The wealthy and powerful, especially rulers, collected animals, sometimes sending expeditions into remote regions for that purpose. They kept them to entertain and impress others, as symbols of their power and the geographical extent of their dominions. Eventually, the animals held in royal menageries became the founding stock for zoological gardens, where they provided entertainment and educational instruction to ordinary people and, more recently, became subjects for zoological research and conservation.

Animals served as more than status symbols. They and their behavior provided moral lessons. This book describes some of the moral tales that were woven around animals in the Middle Ages. The authors of those tales had not visited the lands inhabited by the animals they described, so they had only the accounts of sailors and adventurers to guide them. Not surprisingly, many of their written descriptions and illustrations bore little relation to the actual animals, and some of the animals did not exist at all—indeed, they were biologically impossible. As though descriptions of fantastic animals were not enough, medieval writers also described fantastic people living in distant lands. This book also describes a few of these.

Moralizing myths eventually gave way to more accurate depictions, and the book continues with an outline of their development. It describes how naturalists studied the physiology of animals and the techniques they devised for classifying them. Their work expanded when the invention of the microscope gave them access to the world of the very small, allowing them to watch minute animals and to examine the detailed structure of larger animals.

Animal classification led to theories about the origin of animal species. This led, by the 19th century, to the growing recognition that modern species were the descendents of earlier species that were different in important respects. The book tells how ideas about the evolution of species developed and how the mechanism of heredity was revealed. It describes the discovery of the gene and the physical structure of DNA and how the genetic constitution of species came to be used to determine their relationships.

Finally, the book turns to the study of animal behavior. It describes the work of some of the pioneers in this field and the aspects of behavior they investigated. It tells of the conditioned responses of Pavlov's dogs, the dance of the honeybee, and more.

People have been thinking about and studying animals for thousands of years. In the course of that long history they have accumulated vast amounts of information and, today, when the tools available to scientists are more powerful than they have ever been, information is arriving at an ever-faster rate. Even if it were possible to compress all of that information, or even a comprehensive summary of it, into a single volume, the result would be unwieldy and confusing. *Animals* makes no such pretence—it amounts to nothing more than a series of snapshots providing brief glimpses of some of the events that have led to the present scientific understanding of animals and short accounts of the lives of a few of the remarkable individuals who have contributed to that understanding. The book has a glossary defining the technical terms used in the text, and for readers who would like to pursue the subject further there is a list of books and Web sites where they will be able to learn more.

Where Do Animals Come From?

Animals—or as modern scientists would say nonhuman animals—share the world with us, and they are everywhere. Insects, spiders, and mice take up residence in human dwellings, and no one is ever very far from a rat. People share their homes with companion animals—pets—that have learned to accommodate themselves to human ways. Dogs, first welcomed as hunting partners, have lived with people for about 13,000 years. Cats came indoors about 8,500 years ago, probably when children adopted orphaned kittens that grew up with a taste for the comforts of the fireside and scraps from the table. Previously, they earned their living hunting mice and rats in grain stores close to human dwellings. To a bird, a tall building is equivalent to a precipice, providing nesting sites and, for peregrine falcons (*Falco peregrinus*), perches from which to dive onto prey.

From the earliest times, people have lived with nonhumans and wondered about them. How did they originate? Every culture has a story describing the origin of humans, but what about nonhumans? Throughout history, many people have believed that humans possess souls that continue to exist after the body has died. Do nonhumans have similar souls?

Modern Western science began in ancient Greece. This chapter describes Greek ideas about the origin of animals and compares these with the beliefs of ancient Egypt and of some Native American peoples.

THE ELEMENTS

Scholars accept that Greek philosophy began with a school of thinkers, the first of whom was Thales of Miletus (ca. 624–546 B.C.E.). Miletus was a Greek trading city on the coast of what is now Turkey. Since long before Thales, however, Greeks had pondered and speculated about how the world they saw around them came to be the way it was. They sought to explain things, and to do so they needed to discover the basic substance from which everything is made. Prior to Thales, the Greeks had attributed everything to the actions of supernatural beings. Thales rejected this idea, insisting that natural objects and phenomena had natural, not supernatural, explanations. He was compelled to do so, because supernatural explanations ultimately lead nowhere. If a god fashioned sheep, cattle, and other animals, which god was it? And why did the god make them? How did the god come to live in the sky or the forests in the first place? The questions go on forever, and in the end the answers tell people nothing useful about the animals they are supposed to be describing.

Thales sought explanations for the world. He believed that water was the most fundamental substance of all. He taught that the Earth floats on water and that earthquakes are caused by waves in the cosmic ocean. Anaximander (ca. 610–546 B.C.E.), a follower of Thales, disagreed. Although he accepted that everything originated from a single basic substance, he maintained that that substance could not be water, because water can only be wet, so things made from water cannot be dry. Similarly, fire can only be hot, so cold things cannot be made from fire. He proposed instead that the fundamental substance was what he called the *apeiron,* a word that means indeterminate or boundless. A seed separated from the *apeiron* and grew into the universe, and living things resulted from the action of sunlight shining on water.

Anaximander made another point. If babies appeared without parents, they could not survive. That means there cannot have been a first human or pair of humans from whom all others are descended, because those first humans must have begun life as infants, but infants without parents. They could not have survived (even if they could have come into existence at all). Consequently, he argued, humans must have developed from an earlier, nonhuman form—from an animal.

Other philosophers held that different underlying substances were primary. Anaximenes, who lived around 550 B.C.E., believed

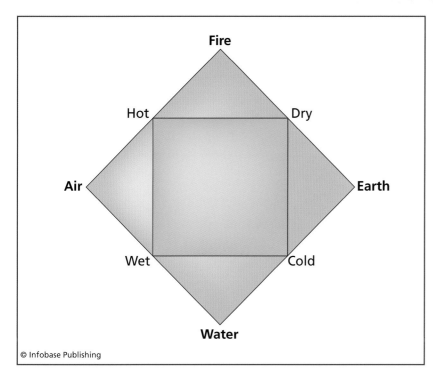

Fire

Hot   Dry

Air   Earth

Wet   Cold

Water

© Infobase Publishing

This traditional representation of the four elements shows one square enclosed by another. The elements—air, earth, water, and fire—are shown at the corners of the larger square and their properties at the corners of the smaller square.

air was the basic substance from which everything is made. A little later, Heraclitus, who lived around 500 B.C.E., believed that the basic substance was fire.

The philosopher Parmenides (ca. 520–ca. 450 B.C.E.) argued that since everything arises from a single underlying substance, information that is based on sensory perception must be illusory, because the original substance constantly changes to produce the objects and phenomena that fill the world. The senses perceive only those objects and phenomena that are products of change, making it impossible to discern the underlying, unifying reality. It is only pure reason that can reveal the truth about the world.

It was Empedocles (ca. 490–430 B.C.E.) who proposed that the underlying substance must form four distinct roots: earth, air, fire, and water. Plato (428 or 427–348 or 347 B.C.E.) called Empedocles' "roots" elements, the name that has endured. The illustration above shows the traditional way the elements and their properties were depicted.

These elements were not what a modern scientist understands by the word. They had nothing in common with the elements listed in the periodic table. Originally they were gods, but Thales and the philosophers who followed him were coming to conceive of them as entirely natural. The English philosopher and mathematician Bertrand Russell (1872–1970) pointed out that at one time physicists believed all matter was derived from hydrogen, by far the most abundant of all the chemical elements in the universe. This was incorrect, but Thales's theory was no more unreasonable.

Thales, Anaximander, Anaximenes, Heraclitus, Parmenides, and Empedocles lived before the time of Socrates (ca. 470–399 B.C.E.), so they are among those ancient Greek philosophers called *pre-Socratic*. Very little is known about their lives, and their teachings are known reliably only from fragments of their writings. Later philosophers described the teachings of individual pre-Socratic philosophers, but mainly in order to disagree with them, so these later accounts are likely to be biased. The pre-Socratics were the first thinkers to attempt to break free from mythological explanations of the origin of the world and all it contains. They were the first rationalists, and Western science began with the attitude they introduced.

EPICURUS AND ANIMALS THAT EMERGE FROM THE GROUND

The ancestors of the Greek philosophers had lived by hunting game and gathering wild plants. Parties of hunters used to leave their families for days or even weeks at a time to seek food in the forests and other wildernesses. There they faced many dangers, not least from wild animals that were more than capable of killing them. The beasts they hunted for food were elusive, fleet of foot, and fierce if wounded or cornered. Probably the hunters believed that the spirits of all of these animals were united in a single goddess known as the Mother of the Animals. Later this goddess came to be known to the Greeks as Artemis and to the Romans as Diana. She carried a bow, quiver, and arrows, or sometimes a spear, she was accompanied by dogs or stags, and her powers were formidable. She killed mortals who offended her. At the same time, she protected children as well as animals. In Asia Minor (now Turkey), Artemis was worshipped as the mother-goddess. In killing game animals, hunters were taking them from their supernatural protector. That was a dangerous thing to do, and the best way to appease the enraged goddess was to return part of what they had taken. So they sacrificed animals to their goddess. That is the origin of religious sacrifices.

The pre-Socratic philosophers were brought up in this tradition. It colored their attitude toward animals, but it did not explain where animals came from. Empedocles had a theory. He believed that portions of the four primary elements mix to form physical objects, and that the characteristics of an object result from the ratio of the primary elements composing it. The coming together and dissolution of the elements are driven by two opposing forces: love and strife. The relative influences of each of these grow and decline in a cycle that Empedocles called the vortex. At one time, love (which Empedocles associated with goodness) was totally dominant and the four elements were joined in a sphere, each element occupying one quarter of the sphere. But then strife (associated with evil) slowly made the sphere disintegrate, scattering the elements randomly throughout the universe. Then love began to reassert itself, drawing the elements together into clusters that formed living organisms. These organisms did not start out as the animals seen in the world today, but as "spare parts." Fingers, toes, knees, heads, and other body parts all formed at random in the ground. They were alive and moved around, and the force of love made them join with such other body parts as they chanced to meet. This process produced animals, but most of them were either grotesquely deformed and incapable of living or unable to reproduce, so they died out. The survivors, possessing bodies that functioned effectively and were capable of reproduction, were the ancestors of all living animals. Life results from love, but in time the continuance of the cycle will make strife more prominent. Then everything will dissolve into a chaotic mix of the primary elements, a state in which everything remains until love becomes reassertive and the organization of matter recommences.

It is tempting to see an early theory of evolution in this, but there are profound differences. Empedocles saw the formation of animals as an evolutionary process, but one that occurred over a particular period in the past. In his philosophy, animals were not still forming or changing their forms.

The philosophical interpretation of reality was slowly changing. At first, there was a single original substance from which everything developed. Then the single substance became the four elements. With Empedocles, the ascendancy of strife destroyed all structure, so there was an infinite number of parts that gradually assembled themselves under the influence of love. This fragmentation led to the theory advanced by Democritus (ca. 460–380 B.C.E.), which held that the universe is composed of atoms. These are indivisible particles—the Greek word *atomos* means indivisible—separated by empty space, and they move downward. Physical forces make atoms move randomly and bring them together to form structures. These structures are temporary, however, and eventually they disintegrate. The philosophers holding this view were called atomists. Their central point was that the universe and everything it contains are formed by natural forces and not by supernatural beings.

Epicurus (341–270 B.C.E.) developed the ideas of Democritus into a major school of atheist philosophy. His followers, the Epicureans, believed that all material things are made from atoms that move downward, but Epicurus differed from Democritus in believing that the falling atoms sometimes swerved to the side. Swerving increased the probability of collisions, and this randomness meant the process of formation was not predetermined. The Epicureans also held that the gods have no influence on human lives, and that the purpose of life was the pursuit of tranquillity. Tranquillity could be found by limiting desires—Epicurus ate only bread, with a little cheese on feast days—and ridding oneself of the fear of gods and death.

If all material things result from atoms joining together, how did the first animals arise? Epicurus believed that the Earth is a mother and that animals arose spontaneously from the ground. When sunshine warmed the ground and showers of rain moistened it, wombs developed, attached to the ground by roots, and when the embryos they contained were sufficiently developed they emerged, rejecting the water in which they had grown and seeking the air.

Epicurus was probably born in an Athenian colony on the island of Samos, in the Aegean Sea. That is where his father lived and where Epicurus was raised. He began to study philosophy when he was about 14. When he was 18, Epicurus moved to Athens, possibly in order to establish his right to Athenian citizenship. In 322 B.C.E., while he was in Athens, the Athenians were expelled from Samos. His family settled as refugees in Asia Minor, where Epicurus joined them and resumed his studies under the followers of Democritus and

Plato. He founded his first school in Asia Minor in 311 B.C.E. and in 307 B.C.E. he moved it to Athens, where he died.

NATURE THE CREATOR

The Epicurean school of philosophy became very influential, but the belief that animals were created by the gods survived its extreme materialism. Stoicism was the chief rival to Epicureanism and contemporaneous with it. The school was founded by Zeno (344–262 B.C.E.), a Phoenician who was born in Citium, a town in Cyprus, but moved to Athens. He first discussed his ideas in the *Stoa Poikile* (painted colonnade) at the agora (main square), which is how his followers came to be called Stoics. At first Zeno's ideas were most popular among Syrians. Later they became influential in Rome, and the school lasted for a very long time and, in fact, it remains influential to this day. When people admire a person's stoic calm under extreme adversity, they are expressing a view taken directly from Stoic teaching.

Zeno was a materialist. He took a commonsense view of the world and had no time for abstractions. He trusted the senses. This was not the whole story, however. The Stoics also believed that the entire course of nature was ordained by a benign Lawgiver, who was sometimes called God and sometimes Zeus. The Lawgiver had designed everything, to the smallest detail, so that events always had natural causes. The Lawgiver was not separate from the world, but infused every part of it, so that all things are components of nature and all things have a purpose connected with humans. For example, some animals are good to eat, some provide leather, feathers, and other useful materials, and others test the courage of those who encounter them. This is a belief that spread beyond Stoicism (see "Moral Lessons to Be Drawn from Animal Behavior" on pages 60–61).

Stoicism teaches that the Lawgiver, or Nature with a capital *N*, has crafted everything that exists with the definite purpose of sustaining life. Animals and humans are equally part of the natural world, and humans should aim to live in harmony with that world.

Ethics are extremely important in Stoic philosophy. Since the Lawgiver has determined what happens in nature, it is impossible for a person to behave unnaturally, but life is good only when the person deliberately pursues ends that are in harmony with natural law. Such qualities as justice, moderation, wisdom, and courage are good, and the Stoic seeks only that which is good. Other things, such as health and wealth, are not good, but they possess value, which makes health preferable to sickness and wealth preferable to poverty. A poor, sick, or oppressed person can be virtuous. There are formidable logical difficulties with this line of ethical reasoning that become evident when it is explained in detail, but over the centuries many people have found it attractive.

Speculation about the origin of animals was not confined to the Greeks, of course. Other cultures had their own explanations, and many people believed that animals preceded humans. The worship of animals was widespread. The Egyptian deities took animal forms, and many are depicted as people with the heads of nonhumans. The relationship was not simple, however, and more than one deity might take the form of a particular animal and certain deities had more than one animal form. Hathor, for instance, was the goddess of love and fertility. She often appeared as a cow or with the head of a cow—as did Isis, the goddess of nature. Horus, the god of light and goodness, had the head of a falcon. Bastet, the goddess of the home, usually had the head of a cat, but sometimes of a lion. Mut, mother of the Moon god, bore the head of a vulture or appeared as a lioness.

In North America, the Caddo people from Louisiana and the Southern Plains believed that at one time there was no distinction between humans and other animals. They all lived together, but as time passed their numbers increased until there was insufficient food. They held a council to decide what to do about it and agreed they would have to separate and some would have to change into nonhuman forms. Thereafter, humans and nonhumans would live apart. Some of the group rolled in the ashes of a prairie fire, which blackened them, and they arose as bears. Others rolled in unburned grass and became buffalo, and other animals followed.

An Eskimo creation myth tells of the Raven, who was born out of the darkness. He set out to discover more of his surroundings and came to realize he was the creator of all life. He grew stronger and flew out of the darkness and found land—the Earth. Wanting living things to populate the Earth, he created plants. Then, as he flew over the land, he saw a giant peapod that burst open, revealing a man—the first Eskimo. The Raven fed the man by creating caribou and musk oxen for him and teaching him how to hunt and how to make a canoe.

The Haida, who live in British Columbia, tell a very similar tale, but in their version the Raven discovers the first people hiding inside a clamshell.

In the Indian tradition, an original solitary being grew bored and lonely, so divided into two. The two were male and female, so they could reproduce. As the generations passed and the numbers of descendants increased, they began to assume different animal forms. In this story all animals are descended from a common ancestor.

ANIMALS DESIGNED FOR THE WAYS THEY LIVE

The cheetah (*Acinonyx jubatus*) hunts on the African plains. It selects a target animal, then stalks it patiently, sometimes for hours, taking advantage of every scrap of cover and taking care to remain downwind of its prey at all times. Finally, when it has crept to within range, it launches its attack. Startled, the prey runs, but no animal can outrun a cheetah over a short distance. The cat tires quickly so the chase is brief, but during it the cheetah is probably capable of more than 60 MPH (96 km/h). It is said to be the fastest of all mammals.

The illustration on page 10 shows how the cheetah achieves such a high speed. Its skeleton is very flexible, allowing its spine to curve both upward and downward, and its pectoral and pelvic girdles allow it to bring its front and back legs so close together that they cross and then extend them almost horizontally. This skeletal flexibility gives the cheetah a hugely long stride, and while it moves at speed its long, strong tail stabilizes the animal. Obviously, the cheetah is superbly equipped for its life as a hunter on the open grassland.

Most bats hunt insects, catching them on the wing. The bat hunts at twilight, when many insects are carried aloft by rising warm air, and it has a sensitive echolocation system that detects prey from a distance and can work in total darkness. The bat is also highly maneuverable, so it can twist and turn through the air following an insect's attempt to escape. The bat is clearly designed for its way of life as a hunter of insects.

The list of similar examples is as long as the list of animal species. Gazelles have good vision for detecting predators and long, strong legs for running away. Birds have wings, which make it possible for them to fly, and anteaters lack teeth, which are not needed by animals that feed exclusively on ants and termites, and long, sticky tongues

© Infobase Publishing

The cheetah (*Acinonyx jubatus*) is built for short bursts of very high speed. Its skeleton is very flexible, giving it a very long stride, and its long tail stabilizes the animal when it is running. At full speed a cheetah can probably achieve more than 60 MPH (96 km/h) over a short distance.

that can wiggle their way through narrow passages and capture the occupants of an ant or termite nest. Every animal is equipped for the life it pursues.

This was the way the Stoics saw the world. The Lawgiver, or Nature, constructed all the animals to perform particular functions. After all, if there were no hunters, the plant-eating animals would multiply until they died miserably from starvation, and, if there were no plant-eaters, the vegetation would choke itself. The craftsmanship of Nature is far better than anything a human worker could achieve, but the difference is only in quality, not principle. It takes a long apprenticeship and years of practice for a worker to make a fine chariot. It is much more difficult to make a cheetah or a bat, but the difference lies only in the necessary level of skill.

The idea that an object is made for a specific purpose is called *teleology,* and the Stoic view of the natural world was teleological. In about 45 B.C.E., the Roman orator Cicero (Marcus Tullius Cicero, 106–43 B.C.E.) wrote a three-volume book with the title *De Natura Deorum* (On the nature of the gods). The book takes the form of a debate between Velleius and Balbus contrasting Stoicism and Epicureanism. Balbus defends Stoicism, and in the second volume he sets out the teleological Stoic position in the following words:

> If, then, the things achieved by nature are more excellent than those achieved by art, and if art produces nothing without making use of intelligence, nature also ought not to be considered destitute of intelligence. If at the sight of a statue or painted picture you know that art has been employed, and from the distant view of the course of a ship feel sure it is made to move by art and intelligence . . . with

what possible consistency can you suppose that the universe which contains these same products of art, and their constructors, and all things, is destitute of forethought and intelligence?

The teleological interpretation of the origin of animals seems persuasive, and it has remained popular throughout history. It survives even today. But modern biologists reject it because they have a more plausible explanation for the way animals are equipped for the way they live. The cheetah's flexible skeleton, the bat's echolocation, the wings of a bird, and every other example of anatomical structures that contribute to animals' success in their environments are instances of *adaptation*—the physiological, behavioral, or inherited characteristics that fit an animal for life in a particular environment (see the sidebar on page 12).

The essential difficulty with teleology is that it makes history run backward. If an animal is formed in such a way as to make it perform a certain role, then the role must be defined before the animal can be designed to fill it. The Greeks believed history was cyclical, so the world and everything in it eventually disintegrates into a chaotic jumble, but if history is continuous, teleology requires starting at the end point and working back in order to design the animals that will fill each role as that role is defined. The future therefore determines the present. This contradicts our common experience, which is of causes preceding their effects.

THE SERVANTS OF PEOPLE

The first animal to be domesticated was probably the wolf (*Canis lupus*) and that domestication occurred in the Middle East. Domestication invariably leads to physiological changes, and by 12,000 years ago the tamed wolf had become a dog (*Canis lupus familiaris*)—a subspecies of the wolf. The jawbone of a domestic dog of that age has been found in a cave in Iraq. Sheep and goats were domesticated in the Middle East between about 9,000 and 7,000 years ago. Cattle were domesticated about 8,000 years ago in what is now Iraq and independently about 7,000 years ago in the Indus Valley, Pakistan. Pigs were domesticated in the Near East about 9,000 years ago. Horses were domesticated in southern Russia about 5,000 years ago.

By the time of the Greek philosophers, the familiar animals of farms and firesides had been domesticated for thousands of years. They were tame, docile, and, in the days before refrigeration when animal products had to be obtained close to where they were consumed, a common sight even in cities such as Athens. There was no reason for anyone to imagine a time prior to their domestication, when their

THE MODERN CONCEPT OF ADAPTATION

Animals can adapt to a change in their surroundings by altering their behavior or seeking a different type of food. That is a form of adaptation similar to acclimatization. In summer people wear light clothes to keep cool and in winter they wear thicker clothes to keep warm. This is an example of a behavioral adaptation to seasonal changes in the weather.

Evolutionarily, adaptation is more permanent. Individual animals that possess some feature that makes them more successful than others at finding food or mates are likely to produce more offspring than rivals lacking that feature. If the offspring inherits the feature it, too, will be successful, and the feature will occur in its descendants. The feature, whatever it may be, allows the animals possessing it to survive better in their environment. It improves their adaptation to their surroundings.

Domestic cats have sharp claws that retract into folds of skin in the paws. Retractable claws allow cats to walk silently across hard surfaces, increasing their efficiency at stalking prey. Retraction also protects the claws from wear and allows cats to climb swiftly in pursuit of prey, and their sharp claws help them grip their prey securely. The possession of retractable claws is an example of adaptation.

Dogs, in contrast, cannot retract their claws, and these are blunt through friction with the ground. They are useless for seizing prey or for climbing, and when a dog walks across a hard surface its footsteps are clearly audible. A dog's narrow paws equip it for running fast, however, and its ability to run down its prey is the dog's adaptation to its way of life.

Woodpeckers have very strong beaks and very long tongues. They use their beaks to open up crevices in the bark of trees and their tongues to extract the insect larvae living beneath the surface. They also use their beaks to excavate holes in trees in which to build their nests. The woodpecker's beak is another instance of adaptation.

Tigers have very conspicuous stripes. At least, their stripes appear conspicuous when the animals are seen in a zoo. In their natural habitat, however, its approximately vertical stripes make a tiger very difficult to see against a background of slender tree trunks and dappled light. The stripes are camouflage, and camouflage is another form of adaptation.

Every living organism possesses features that adapt it to its environment. It is not always simple to identify those features that are genuinely adaptive, but adaptation is a consequence of natural selection. It is the way evolution proceeds.

ancestors had lived as wild animals. Consequently, they fitted well into the teleological system of the Stoics. Oxen have strong necks and shoulders that are obviously intended to wear yokes attached to plows and wagons. Sheep have fleeces of wool from which cloth can be woven and they provide milk. Clearly, that is their purpose. Every domesticated animal has a purpose, so to the Stoics it followed that each animal had been made specifically for that purpose.

They carried the idea further. The Stoics believed that the purpose of rain is to provide drinking water and irrigate crops. Wide rivers exist to be navigated. Ultimately, everything that exists has a purpose that is in some way related to humans and meant to serve them.

It is easy to see how the Greeks might suppose that domesticated animals were made specifically to serve humans. Domestication commences when people adopt animals they have managed to reassure sufficiently for them to be handled. Children often adopt as pets young animals such as puppies, kittens, and lambs. When they grow up, they are accustomed to human company and are not frightened of the people whose homes they share. Larger animals such as cattle would have been corralled at night to protect them from predators, and farmers would have brought into the coral those individuals that allowed themselves to be handled and even milked. With each generation the farmers selected for breeding the animals with the most desirable qualities—those that grew biggest or produced the most milk or wool and that were easy and safe to manage. Offspring inherited their parents' desirable characteristics and over many generations these became more pronounced. Farm animals grew more docile and more productive. Selective breeding can lead to extremes of difference: Every breed of domestic dog, from the Chihuahua to the Great Dane, from the toy poodle to the Rottweiler, is descended from the wolf that was the ancestor of all domestic dogs. After thousands of years of domestication and selective breeding, farm animals were indeed the servants of humans, but only because humans had made them so.

The Stoics were not alone in believing that domestic animals were made to be servants. This belief also occurs in the Judaic and Christian tradition that God has given people dominion over nature. In Genesis (1:28–29), God instructed the first man and woman to: "Be fruitful, and multiply, and replenish the earth, and subdue it; and have dominion over the fish of the sea, and over the fowl of the air, and over every living thing that moveth upon the earth."

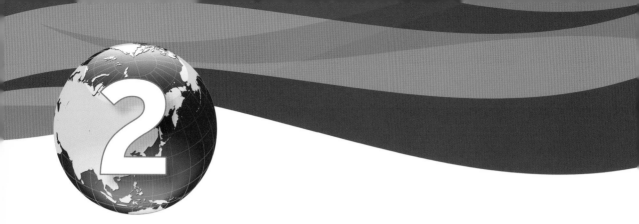

The Beginning of Science

The farmers who domesticated animals did not set out with a plan. There was no genius who one day proposed that they should establish a captive breeding program with the aim of developing beasts that would fulfil particular functions. It just happened that when farmers sought to protect cattle and sheep from wolves and other predators they inadvertently began a selective breeding program that led in time to full domestication.

Nor did the pre-Socratic philosophers base their understanding of the world on close observation and experimentation. Their view was not scientific, and neither were the views of the Epicureans or the Stoics. Empedocles (see "Epicurus and Animals That Emerge from the Ground" on pages 4–7) did not discover the evolution of species some 2,000 years ahead of the biologists of the 19th century, despite the superficial similarity between his theory and the scientific one. Democritus and the atomists who followed him believed the universe to be composed of indivisible particles they called atoms. Similar theories were propounded in India in the sixth century B.C.E. The atomists and their Indian counterparts are often said to have stumbled on a correct view of the nature of matter that was not confirmed until the discoveries made by the English chemist, physicist, and meteorologist John Dalton (1766–1844) early in the 19th century. But the Greek and Indian scholars understood their atoms and the properties of those atoms very differently. The only feature they share with Dalton's theory is that they describe an ultimate component of matter than cannot be subdivided into smaller parts—and even that similarity evaporated with the discovery of subatomic particles.

A different way of exploring the world began to emerge in the fourth century B.C.E. This chapter briefly outlines that emergence into what would, many centuries later, become the scientific approach. The chapter begins with two schools of Athenian philosophers, continues with one of the most famous encyclopedias of natural history ever published, and ends with one of the most famous of all volcanic eruptions.

PLATO AND THE WORLD OF IDEAS

Parmenides (see "The Elements" on pages 2–4) believed that our senses deceive us by exposing us to illusions while concealing the underlying reality. This strongly influenced Western history's second most famous philosopher, Plato (428 or 427–348 or 347 B.C.E.). The most famous of all philosophers was Socrates (ca. 470–399 B.C.E.). Plato studied under Socrates and admired and loved his teacher.

Plato was born and died in Athens. A mathematician, sociologist, and political theorist, he founded the Academy in Athens in about 387 B.C.E. and presided over it until his death. The Academy was a kind of debating society. It was not open to the public, although it charged its members no fees.

Plato was an aristocrat related to people linked to the tyrants who ruled Athens, and he lived in troubled times when Athens was at war. Two of the tyrants were his uncles. Plato bitterly hated democracy, a hatred that is clearly expressed in the 10 books of his *Republic.* In the *Republic,* written in the form of a dialogue in which Socrates explains matters to his friend Glaucon, Plato describes the political structure of what he regards as the ideal form of government: It is a totalitarian dictatorship. The book has been highly influential throughout history. In his *History of Western Philosophy* Bertrand Russell wrote: "Why did the Puritans object to the music and painting and gorgeous ritual of the Catholic Church? You will find the answer in the tenth book of the *Republic.* Why are children in school compelled to learn arithmetic? The reasons are given in the seventh book."

Troubled times are times of rapid change, so perhaps it is not surprising that Plato believed that all change was for the worse. He sought permanence but found only change, and in his pursuit of permanence he developed the ideas of Parmenides, expounding them in books V to VII of the *Republic.* Plato deals first with the meaning of words. For example, everyone knows what a dog is, but if someone says *dog* the word might suggest a terrier to one person, a spaniel to another, and a St. Bernard to a third. The word *table* describes many different kinds of objects, all of which are tables. But what is it, Plato asks, that all these different dogs and tables have in common? Clearly, it is some quality of dogness or tableness, some basic, defining characteristic. If someone points to an actual dog, there was a time before that dog was born. It has not always existed. The same is true of a particular table, which must have been made by someone at some past time. Dogs and tables come and go, and the same is true of everything else in the world. Change is everywhere. But dogness and tableness are always there in the background as the abstract qualities that make it possible to identify objects. There is the permanence Plato sought.

The argument then moves away from Parmenides and becomes metaphysical. *Metaphysics* is the branch of philosophy concerned with ideas of being, knowing, and the mind. Plato maintains that words describing objects must refer to original models—archetypal examples of those objects created by God. So all dogs and tables can be identified by reference to the archetypes, which have the essential qualities all dogs and tables possess. This applies not only to objects but also to qualities such as colors.

The archetypes are ideas or forms, and this part of Plato's philosophy is called the theory of forms. Its central point is that only the forms are real. The particular examples of the forms that people see, hear, touch, taste, or smell are only apparent. They are illusory. Plato explained his theory of forms in the following allegory (in the translation by Benjamin Jowett) at the beginning of book seven:

> And now, I said, let me show in a figure how far our nature is enlightened or unenlightened:—Behold! human beings living in an underground den, which has a mouth open towards the light and reaching all along the den; here they have been from their childhood, and have their legs and necks chained so that they cannot move, and can only see before them, being prevented by the chains from turning their heads. Above and behind them a fire is blazing at a distance, and between the fire and the prisoners there is a raised way; and you will see, if you look,

a low wall built along the way, like the screen which marionette players have in front of them, over which they show the puppets.

I see.

And do you see, I said, men passing along the wall carrying all sorts of vessels, and statues and figures of animals made of wood and stone and various materials, which appear over the wall? Some of them are talking, others silent.

You have shown me a strange image, and they are strange prisoners.

Like ourselves, I replied; and they see only their own shadows, or the shadows of one another, which the fire throws on the opposite wall of the cave?

True, he said; how could they see anything but the shadows if they were never allowed to move their heads?

And of the objects which are being carried in like manner they would only see the shadows?

Yes, he said.

And if they were able to converse with one another, would they not suppose that they were naming what was actually before them?

Very true.

And suppose further that the prison had an echo which came from the other side, would they not be sure to fancy when one of the passers-by spoke that the voice which they heard came from the passing shadow?

No question, he replied.

To them, I said, the truth would be literally nothing but the shadows of the images.

That is certain.

The allegory of the prisoners in the cave is meant to show that the world we see around us is but a shadow of the real world of forms. There is no way anyone could investigate the theory of forms, so it cannot be proved or disproved and therefore is not a scientific theory. It has had one interesting consequence, however, and one that reflects exactly Plato's desire for permanence. If the objects and qualities in the world are illusory and each refers to its form, which is the only reality, then species of plants and animals are fixed. The cheetah that hunts on the plains is an illusion, but God has made a real and unchanging cheetah that stands for all cheetahs, and the same relationship applies to every species. Animals, in Plato's philosophy, cannot evolve.

ARISTOTLE AND THE IMPORTANCE OF OBSERVATION

Plato's Academy, located in an Athenian olive grove, produced many scholars. The most famous of them by far was Aristotle (384–322 B.C.E.). He was born at Stagirus, a Greek colony on the Macedonian coast, the son of Nichomachus, the personal physician to Amyntas III, the king of Macedonia. His father died while Aristotle was a boy, and he was raised by a guardian. In about 367 B.C.E., when he was 17, Aristotle joined Plato's Academy and remained there for 20 years as a pupil and later as a teacher. He left after Plato's death and spent time in Anatolia and the island of Lesbos before returning to Macedonia in 343 B.C.E. Amyntas had died, and the new king, Philip II, appointed Aristotle as tutor to his son Alexander, who later became Alexander the Great. Aristotle returned to Athens in about 335 B.C.E., where for the next 12 years he taught at the Lyceum, a school established in the grounds of the temple to Apollo Lyceius, hence its name. Alexander died in 323 B.C.E., and those who opposed the Macedonians became powerful in Athens. Aristotle's name was linked to that of a Macedonian general, and he was charged with impiety—a capital offence. He avoided trial by moving to Chalcis (modern Khalkis) on the island of Euboea, north of Athens, which is where he died the following year.

Aristotle had many interests, but he devoted a great deal of his time to natural history. He established a zoo at the Lyceum, stocking it with animals captured during Alexander's Asian campaigns, and he studied the natural history of Lesbos as well as of adjacent areas and the surrounding seas. Aristotle broke from the Platonic way of thinking. Plato had emphasized the primacy of ideas over observation, which he considered illusory. Aristotle emphasized careful observation. His studies of animals involved dissection. He broke open chicken eggs at intervals between fertilization and hatching and recorded the order in which organs developed. He recognized that sharks and rays were closely related, and he distinguished between aquatic mammals and fishes.

In the nine books of his work *The History of Animals*, Aristotle described more than 500 species, including 120 species of fish and 60 species of insects. He distinguished between vertebrates and invertebrates, although he called them "sanguinea" (animals with blood) and "animals without blood." He divided the animals with blood into those that bear live young and those that lay eggs. His "animals without blood" included insects, mollusks, and crustaceans with and without shells. Aristotle also wrote *On the Generation of Animals,* in five books. This work describes all the different ways in which animals reproduce and goes into the details of reproductive anatomy.

Although Aristotle relied on observation, the conclusions he drew were often mistaken. In the third book, in trying to explain why some birds lay more eggs than others, for instance, he had the following to say about the cuckoo:

> The cuckoo, though not a bird of prey, lays few eggs, because it is of a cold nature, as is shown by the cowardice of the bird, whereas a generative animal should be hot and moist. That it is cowardly is plain, for it is pursued by all the birds and lays eggs in the nests of others.

A more sympathetic commentator might interpret fleeing from pursuers and hiding its eggs in the nests of other birds as the behavior of a victim rather than a coward! In any case it is wrong, of course, for the cuckoo is a highly successful brood parasite, which is why other birds try to drive it away from their nests. Aristotle realized that animals cannot be classified on the basis of their reproductive method, as the following passage from the second book illustrates.

> Not all bipeds are viviparous (for birds are oviparous), nor are they all oviparous (for man is viviparous), nor are all quadrupeds oviparous (for horses, cattle, and countless others are viviparous), nor are they all viviparous (for lizards, crocodiles, and many others lay eggs). Nor does the presence or absence of feet make the difference between them, for not only are some footless animals viviparous, as vipers and the cartilaginous fishes, while others are oviparous, as the other fishes and serpents, but also among those which have feet

many are oviparous and many viviparous, as the quadrupeds above mentioned. And some which have feet, as man, and some which have not, as the whale and dolphin, are internally viviparous.

Aristotle distinguished animals by a moral standard based on the ratio of the four elements in them:

> ... it is those animals which are more perfect in their nature and participate in a purer element which are viviparous, for nothing is internally viviparous unless it receive and breathe out air. But the more perfect are those which are hotter in their nature and have more moisture and are not earthy in their composition.

It was not until the 19th century that biologists confirmed certain of Aristotle's observations. For example, male cephalopods (octopuses, squid, cuttlefish, and nautilus) possess a specialized tentacle called a *hectocotylus* that is used to store sperm and transfer it to the female. During mating the hectocotylus sometimes breaks off and males usually grow a new one for each mating season. Aristotle described the hectocotylus, probably relying on information from local fishermen. Naturalists dismissed the idea until the organ was rediscovered and named by the French zoologist Georges Cuvier (1769–1832). Aristotle also observed that in a species of dogfish known as the dusky smooth-hound (*Mustelus canis*), the young develop inside the mother's body attached to a structure resembling a mammalian placenta—in fact, a yolk sac. Again, naturalists dismissed the idea until 1842, when the German physiologist Johannes Peter Müller (1801–58) verified it.

Aristotle seems to have possessed boundless energy, and his contribution to zoology was huge. Apart from his own work, he provided guidance for his pupils in the form of lectures and encouragement to undertake research. His classification of animals was often highly perceptive, but his aim was to seek the most fundamental characteristics, of which other characteristics were the consequences. This allowed him to arrange animals into a great chain of being, with those possessing the most basic features at the bottom and other animals ranged above them in levels of increasing complexity. The chain comprised an unbroken sequence of animals, and it united every species. The diagram opposite shows how Aristotle arranged animals. His chain of being remained popular until it was finally abandoned

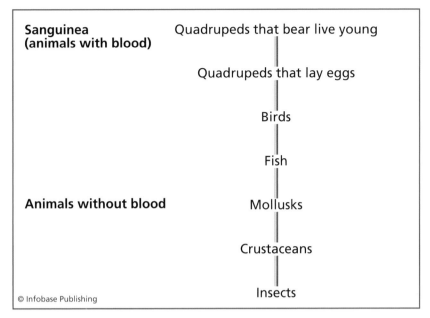

Sanguinea (animals with blood)	Quadrupeds that bear live young
	Quadrupeds that lay eggs
	Birds
	Fish
Animals without blood	Mollusks
	Crustaceans
	Insects

© Infobase Publishing

Aristotle arranged animals into a hierarchy with the most fundamental at the bottom and the most complex at the top. This was the origin of the Great Chain of Being that became increasingly elaborate from medieval times until the development of evolutionary theory in the 19th century led to it being finally abandoned.

in the 19th century as evolutionary theory rendered it irrelevant—as well as incorrect.

Aristotle accepted that every organism must perform a particular function, which meant that its basic features were usually fixed, but he did not accept Plato's idea that each species reflects an idealized form. Species could change, although their capacity to do so was limited by the fact that offspring resemble their parents. He accepted that animals existed to serve humans, but in his day—and for long afterward—no one questioned that.

PLINY, COLLECTOR OF INFORMATION

Scholars continued to gather information in the centuries that followed Aristotle, and many wrote books describing their work. However, by the first century C.E., the impetus was being lost and the rate of new discovery and observation was slowing. Gaius Plinius Secundus (23–79 C.E.), better known as Pliny the Elder, a Roman soldier, government official, and author, feared that the old knowledge might be lost, and he set himself the task of gathering as much of it as he could and recording it in a single, encyclopedic work. That work was called *Naturalis Historia* (*Natural History*), and it was almost finished in 77 C.E. Pliny's adopted nephew, also called Pliny (Pliny the Younger), completed it after his uncle's death. Pliny's *Natural History* comprises 37 volumes covering the physical world, geography, anthropology, human physiology, botany, agriculture, pharmacology, mineralogy, mining, precious stones, sculpture, and more. It was a truly immense work. Volumes VIII to XI were devoted to zoology.

Pliny was born at Novum Comum (modern Como) in northern Italy. He had a sister and his father was wealthy. By 35 C.E. Pliny had arrived in Rome, accompanied by his father, and it is in that city that he was educated and received the military training customary to young men of his social rank. He began his military career in Germania, the region under Roman rule bordering the River Rhine, serving under his father's friend Publius Pomponius Secundus, who was famous as a tragic poet in addition to being an army general. Pomponius inspired Pliny with a lifelong love of learning and used his social and political connections to further Pliny's career. Pliny rose to the rank of cavalry commander and during one winter wrote about the use of lances hurled from horseback. During the reign of the emperor Nero, from 54–68 C.E., Pliny spent most of his time in Rome, some of it working on a history of the German wars. In 69 C.E., Vespasian succeeded Nero, and he appointed Pliny to important official positions that took him to many European parts of the empire. At the end of this period the emperor made Pliny prefect—admiral—of one of Rome's two navies, responsible for the western Mediterranean and based at Misenum on the Gulf of Naples. The map on page 23 shows the location of Misenum and its proximity to Mount Vesuvius, an active volcano. Vesuvius erupted in 79 C.E., and Pliny lost his life during the eruption (see the sidebar on page 25).

In addition to his official duties, Pliny was a prolific author. He was also a workaholic. He slept little and, according to his nephew, the only time he took off from work was while he was bathing, and even then a book was read to him while he was being dried and he was dictating notes on it. When traveling around Rome he had

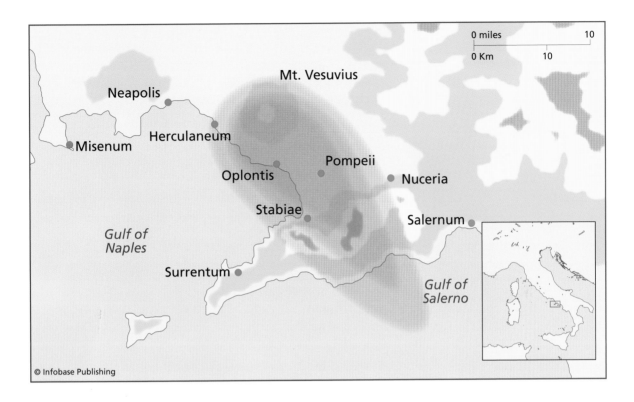

The Gulf of Naples (Neapolis) as it was in 79 C.E. when Mount Vesuvius erupted. The shadow shows the area that was covered in volcanic ash.

himself carried in a chair so he need not interrupt his work, and he scolded his nephew for wasting time walking. Not surprisingly, perhaps, Pliny never married.

All of Pliny's works are lost except for the *Natural History*. That work ends with the words: "Greetings, Nature, mother of all creation, show me your favor in that I alone of Rome's citizens have praised you in all your aspects." Pliny was seeking to claim natural philosophy from the Greeks and to identify it with Roman culture. The result is a truly monumental work. Pliny gives many of the sources of his information and discusses the opinions of other authors, sometimes disagreeing with them. He describes land animals in book 8, marine animals in book 9, birds in book 10, and insects in book 11. The following extract from chapter 10 of book 11, describing the behavior of bees, provides a flavor of Pliny's style.

The manner in which bees carry out their work is as follows. In the daytime a guard is stationed at the entrance of the hive, like the sentries in a camp. At night they take their rest until the morning, when one of them awakes the rest with a humming noise, repeated twice or thrice, just as though it were sounding a trumpet. They then take their flight in a body, if the day is likely to turn out fine; for they have the gift of foreknowing wind and rain, and in such case will keep close within their dwellings. On the other hand, when the weather is fine—and this, too, they have the power of foreknowing—the swarm issues forth, and at once applies itself to its work, some loading their legs from the flowers, while others fill their mouths with water, and charge the downy surface of their bodies with drops of liquid. Those among them that are young go forth to their labors, and collect the materials already mentioned, while those that are more aged stay within the hives and work.

Pliny derived his information from a total of 146 Roman and 327 foreign authors. He was a collector of information rather than an investigator, but where he could he did try to check information. He was a Stoic, believing that natural phenomena have natural explanations, and he declared that he did not believe in immortality. But he had a fascination for the exotic and tended to repeat travelers' tales. These included tales about unicorns and men whose feet pointed the wrong way.

The volcanic eruption that killed Pliny buried the cities of Pompeii and Herculaneum beneath volcanic ash. The destruction was sudden. The eruption occurred in two phases. The first hurled a column of smoke, ash, and pumice high into the atmosphere. That phase lasted for 18 to 20 hours. Then came a *pyroclastic flow*—a cloud of hot ash, carrying pumice and other rocks, that swept down the side of the volcano faster than a man could run. No one caught in its path could survive the heat or the inhalation of volcanic ash, which is powdered rock that sets like cement when mixed with the moisture in the respiratory passages. Dreadful though it was, the ash buried the two cities so quickly that it preserved them. Modern archaeological excavations have now exposed them, revealing streets and buildings that Pliny would have known well. The photograph on page 26 shows one street in Pompeii and the remains of the buildings on either side of it.

Pliny's *Natural History* was so vast and appeared so authoritative that it became a principal source of information. During the European Middle Ages, when monasteries were important seats of learning, many of the larger ones possessed a copy. It was not until the 15th century that its accuracy was questioned. *Natural History* was one of the first books to be printed, and by that time it had been copied by hand so many times that the first and second printed editions were full of obvious mistakes. In 1493, the scholar Ermolao Barbaro (1454–93) published a version in which he claimed to have corrected 5,000 errors, but he was not criticizing Pliny, only the copiers. Niccoló Leoniceno (1428–1524), a physician and professor of moral philosophy, went much further. He criticized Pliny's accounts

THE DEATH OF PLINY

Pliny the Elder was an officer in the Roman army, and at the age of 36 he was made a procurator—an official responsible for collecting taxes, administering money, and in some cases dispensing justice. His duties kept Pliny traveling, and he was seldom in Rome. After some years as a procurator, he was also made a prefect—admiral in modern terms—of the Roman fleet stationed at Misenum on the Bay of Naples. The fleet was responsible for security throughout the western half of the Mediterranean.

In August of 79 C.E., Pliny was at Misenum as usual and his sister Plinia was staying with him together with her son, Pliny the Younger. On August 24, Vesuvius, the large volcano on the opposite side of the bay, became active. Pliny had been out, and on his return home he took a bath. It was after his bath that Plinia drew his attention to the cloud above the volcano. Realizing that people were in danger, Pliny ordered the warships to be launched with the intention of using them to evacuate the inhabitants of the towns across the bay. By the time they arrived it was evening. They landed at Stabiae, where Pliny spent the night with his friend Pompnianus. According to his nephew's account, Pliny dined cheerfully, or

with the pretence of cheerfulness, in order not to alarm his hosts, and then he went to bed.

In the middle of the night Pliny was roused from his bed. Rocks were falling close to the house, and the building itself was shaking badly. Everyone decided they would be safer in the open so they left the house, using pillows to protect their heads from falling stones. By this time it should have been daylight, but the volcanic cloud made it darker than the darkest night. They and other parties also seeking safety had lamps and torches that helped them find their way.

They all decided to go to the shore to see whether they could find safety by sailing out from the coast, but the sea was too rough for them to launch boats. Pliny, who may have been asthmatic, lay down on a linen cloth. Twice he asked for cold water, which he drank. Then they saw flames approaching, and there was a smell of sulfur. Two slaves helped Pliny struggle to his feet, but he collapsed at once. He was inhaling ash and found breathing difficult and painful. The situation on the shore must have been chaotic, for Pliny was not found until the morning of August 26. His body was intact, and he looked as though he was asleep.

In 79 C.E. the Roman city of Pompeii was buried beneath ash from the eruption of Mount Vesuvius. The eruption killed the inhabitants but preserved the city. Archaeological excavations have since exposed it. This is the Via di Stabia. There are stepping-stones to help pedestrians cross the street, and there was a fountain at the intersection in the foreground. *(Erich Lessing/Art Resource, NY)*

of medical herbs in a pamphlet he published in 1492 at Ferrara. When others sought to defend Pliny, Leoniceno went into much more detail, publishing an enlarged edition in 1509. First, he pointed out that Pliny had made the ridiculous claim that the Moon is larger than the Sun. He then claimed that Pliny simply quoted sources without gaining firsthand experience of the topics he described, and this had led him into confusing Latin and Greek names for plants of which he was personally ignorant. Pliny's work continued to influence poets and the authors of nonscientific works, but his influence among natural historians declined, and by 1700 scientists had abandoned it.

Menageries and Zoos

Zoological gardens—zoos—attract millions of visitors every year, and most large cities have one. In rural areas where there is more space, safari parks hold animals in large enclosures crossed by unfenced roads, allowing visitors a view from their cars of animals in more natural surroundings. Modern zoos and safari parks exist primarily to conserve the animal species in their care, some of which are endangered in the wild. Zoologists study the behavior and physiology of zoo animals, and veterinarians watch over their health. Behind the scenes, larger zoos have research and educational facilities. The entrance fees that visitors pay to see the animals helps toward the cost of what is a very expensive operation. Most of the animals have been bred in captivity, and zoos collaborate in breeding programs and exchange animals. Even so, many people think it is cruel to confine animals that have never been domesticated and that are not tame, despite spending their entire lives in captivity.

Conservation is a modern concept. Historically, zoos have served quite different purposes: principally, the study of exotic animals and the display of wealth and power. Kings and emperors have always needed to impress their courtiers—who might seek to usurp them—and visiting foreigners—whose own monarchs might seek to invade and overthrow them. Large, richly furnished and orna-mented palaces advertised the wealth of their owners, who clearly could afford the cost of a war to protect their frontiers. No matter how ornate, however, or how lavish its decoration or the dress of its occupants a palace could not convey an idea of the geographic extent of its owner's dominions. The ruler of a prosperous city might live in a palace that was in every way the equal of the palace of an emperor who ruled half the world.

Exotic animals provided one way to demonstrate military and economic power, and throughout history rulers have collected such animals, brought home from distant lands by their conquering armies. Invited guests—for such royal collections were seldom open to the public—might then be shown animals renowned for their size or ferocity or strangeness. The guests would be agreeably entertained, and they would understand the underlying message. Rulers accepted gifts of rare animals from other monarchs or as tribute from those they had defeated, and they also dispatched expeditions for the express purpose of collecting animals. King Shulgi (2094–2047 B.C.E.) of Ur, in what is now Iraq, owned what was probably the first collection to include large carnivorous animals such as lions and leopards. Natural historians were usually allowed to study the animals in royal collections. Aristotle based his knowledge of zoology largely on the animals collected during the Asian campaigns of Alexander and held at the Lyceum (see "Aristotle and the Importance of Observation" on pages 18–21).

This chapter tells the story of zoos and menageries, and it explains the difference between them. It begins in Asia, continues in Egypt, and then moves to Greece and Rome, where wild animals found other employment as combatants in gladiatorial contests.

KING WEN AND HIS GARDEN OF INTELLIGENCE

Cheng Tang (1675–1646 B.C.E.), the first king of the Shang (or Yin) dynasty that ruled northeastern China, ordered the establishment of a reserve to hold wild animals. In doing so he started a tradition that continued throughout Chinese ancient history. Each king, or later emperor when Chinese control extended more widely, had a menagerie (see the sidebar "The Difference between a Menagerie and a Zoo" on page 45). It was Zhou Wen Wang (1099–1050 B.C.E.), however, who assembled the most famous of these early collections.

King Wen was the founder of the Zhou dynasty, which controlled the valley of the River Wei, near the Yellow River in what is now Shaanxi Province, and he had plans to overthrow the cruel Shang dynasty, but died before he could realize them. Zi Zhou, the last Shang ruler, captured and imprisoned Wen and held him for three years. According to tradition, Wen spent his time in captivity writing the *I-Ching* or *Yijing* (*Book of Changes*), a Confucian classic. He was released when his own people paid a ransom for him. He was regarded as an intelligent, peace-loving, and benevolent ruler. Following his release, he did his best to reform the system of government by making it less cruel and corrupt.

Wen's zoo covered 1,500 acres (607 ha). He called it the Ling-Yu, which means Garden of Intelligence, or Garden for the Encouragement of Knowledge. His garden was walled and it contained deer, antelope, and birds that were possibly peafowl, as well as a pond filled with fish. Visitors were meant to contemplate the animals and consider philosophical questions concerning them. The garden had its own keepers and persons to provide veterinary care. Wen's zoological garden started a fashion, and throughout the Zhou dynasty wealthy families maintained large zoos modelled on the Ling-Yu. This became a tradition and in later times Chinese emperors always kept large enclosed parks containing wild animals (see "Marco Polo and the Gardens of Kinsai" and "Marco Polo at Xanadu" below and on pages 34–36). In 1417 C.E., the Ming emperor Yung-lo (1360–1424) sent an expedition to Africa to capture a giraffe for his collection that already included a zebra and some other African animals.

The Chinese tradition did not extend to Japan, however. There, the first zoo opened in 1725, when a pair of Asian elephants entered the country at Nagasaki. The male was taken to Edo (now called Tokyo) and shown to the public. A small menagerie opened in Edo in 1873. This moved to Ueno Park in 1882, where it developed to become the present Ueno Zoological Gardens. Until Kyoto Zoo opened in 1903, Ueno Zoo was the only zoo in Japan.

MARCO POLO AND THE GARDENS OF KINSAI

In the late 13th century, the Venetian merchant Marco Polo (1254–1324) visited a large city in China that he called Kinsai (or Quinsay). Today, it is called Hangzhou and is the capital of Zhejiang Province in the People's Republic of China. Since 1127, Hangzhou—then called Lin'an—was the capital of the empire of the Song dynasty (960–1279 C.E.). By the time Marco Polo reached it, the national capital had

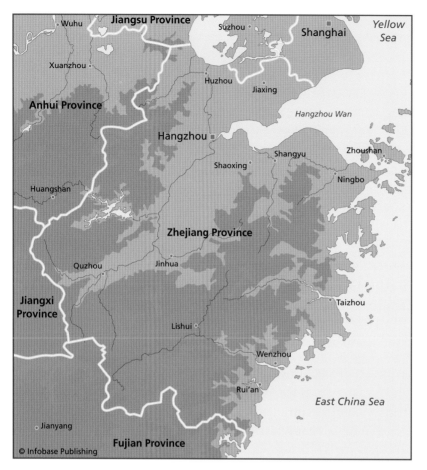
© Infobase Publishing

Hangzhou is the capital of Zhejiang Province of the People's Republic of China. The map shows its location at the head of the Yangtze Delta, beside the Qiantang River. Marco Polo called the city Kinsai. Much of the city that Polo saw was destroyed during a rebellion in the 19th century.

moved to Beijing, and Kublai Khan (1215–94) ruled the empire. The map shows the location of Hangzhou, 112 miles (180 km) southwest of Shanghai.

According to Marco Polo's account, Kinsai contained 1.6 million houses, including many palatial mansions. When he arrived, agents of Kublai Khan were conducting a census, and Polo obtained his information from them. Modern estimates suggest the total population was really 1 to 1.5 million. It is possible that the khan's agents were inflating the population figures to increase the amount of tax the city could be charged. Nevertheless, Kinsai, or Hangzhou, was, and is still, a big and important commercial city.

China had earlier been divided in two. Until Kublai united it into a single empire, the southern part, ruled by Song dynasty kings, had been called Manzi. The last Song king had departed, but Kinsai still contained his palace and its garden. Everything about Kinsai impressed Polo, but especially the palace. "No words of mine could describe its superlative magnificence," he wrote. He stated that the palace and its grounds occupied a space about 10 miles (16 km) in circumference, originally enclosed by high walls topped with battlements. The area was divided into three parts. The first comprised a large and luxurious pavilion, the second a cloistered courtyard containing the royal chambers. A covered corridor, six paces wide, led from this courtyard to a large lake (which is still one of the city's main tourist attractions). There were 10 more courtyards on either side of the main corridor. Each of these 20 courtyards contained 50 houses and each house had a garden. The houses accommodated the 1,000 young women in the service of the king, some of whom accompanied the king and queen when they sailed on the lake in elaborately decorated barges. The main pavilion and other accommodations occupied one-third of the total area. Polo described the third section, two-thirds of the total area—about five square miles (13 km^2) if Polo's estimate was correct—in the following words:

> The other two-thirds of the enclosure were laid out in lakes filled with fish and groves and exquisite gardens, planted with every conceivable variety of fruit-tree and stocked with all sorts of animals such as roebuck, harts, stags, hares, and rabbits. Here the king would roam at pleasure with his damsels, partly in carriages, partly on horseback, and no man ever intruded. He would have the damsels hunt with hounds and give chase to the animals.

Marco Polo did not witness this hunt himself, of course. He writes that he is repeating the account given to him by a merchant he met in Kinsai, "a man of ripe years who had been intimately acquainted with King Facfur and knew all about his life and had seen the palace in all its glory." His new acquaintance took Polo through the grounds, where he reported that "the chambers of the damsels have all fallen in ruin, and only vestiges of them remain. Likewise the wall that encircled the groves and gardens has fallen to the ground, and neither trees nor animals are left."

Facfur was the last Song ruler of Manzi and, says Marco Polo, it was his lifestyle that brought him down. "Amid this perpetual dalliance he idled his time away without once hearing the name of arms. And this at last was his undoing, because through his unmanliness and self-indulgence he was deprived by the Great Khan of all his state to his utter shame and disgrace."

Marco Polo may have exaggerated, but it is probable that the Song kings enjoyed a garden of this kind. Many wealthy individuals maintained grounds of this type, if not of this size.

Marco Polo was born in Venice on September 15, 1254. His father, Niccoló Polo, was a merchant. Niccoló and his two brothers Maffeo and Marco were business partners trading with Asia. In 1259 Niccoló and Maffeo sailed from Constantinople (now Istanbul), where they had been living for several years, to Sudak, a port in the Crimea where there was a Venetian colony and where their brother Marco owned a house. They were seeking better markets and after a time they moved eastward from Sudak, forming or joining a caravan transporting goods to Sarai Batu, a city on the River Volga where Berke Khan, ruler of the Mongol state known as the Golden Horde, had established his capital. They remained there for about a year, but when they wished to return home they found their route blocked by a war that had broken out between Berke and his cousin Hülegü Khan. The Polo brothers sought refuge in Bukhara, in modern Uzbekistan, which was neutral. Stranded in Bukhara, the brothers learned the Mongol language, and when they were offered an opportunity to join a diplomatic mission from Hulagu to his brother, Kublai Khan, they gladly accepted. In 1266, the Polo brothers reached the city of Dadu (modern Beijing), Kublai's capital. Kublai was an intelligent, humanitarian, highly cultured man. He welcomed the two Westerners warmly and listened to all they had to tell him about the West. Kublai sent them back to Europe with a safe conduct—a kind of passport—as well as gifts for the pope and a letter requesting 100 men to teach his people about Christianity and Western customs and some holy oil from Jerusalem. The Polos reached Acre, in what is now northern Israel, in 1269, and from there returned to Italy. Pope Clement IV had died in 1268, and it was 1271 before Gregory X was elected to succeed him. The new pope received Kublai's letter and gifts and blessed the Polos. He could not find 100 scholars, but he sent the oil and two Dominican monks. The monks gave up, but the two Polo brothers, this time accompanied by the 17-year-old Marco, persevered and in 1274 they reached Dadu and presented Kublai with the gifts sent by the pope. Marco was an excellent linguist and entertaining storyteller, and he became a favorite with Kublai Khan.

The three Polos remained at the court of Kublai Khan for 17 years, during which time the khan sent Marco on various missions around the country. In 1291 Kublai released them and they set out for home, escorting a Mongol princess who was to be the bride of Arghun Khan, the Mongol ruler of the Levant. The journey took them two years, sailing through the Malay Strait, around India to Hormuz at the southern end of the Persian Gulf, overland to Trebizond on the Black Sea coast, and by sea from there to Venice. Chinese and Persian sources attest that Marco's sketchy account of the voyage is substantially accurate, although they do not confirm the heroic part the Polos played in their adventures. The map shows the route the Polos followed.

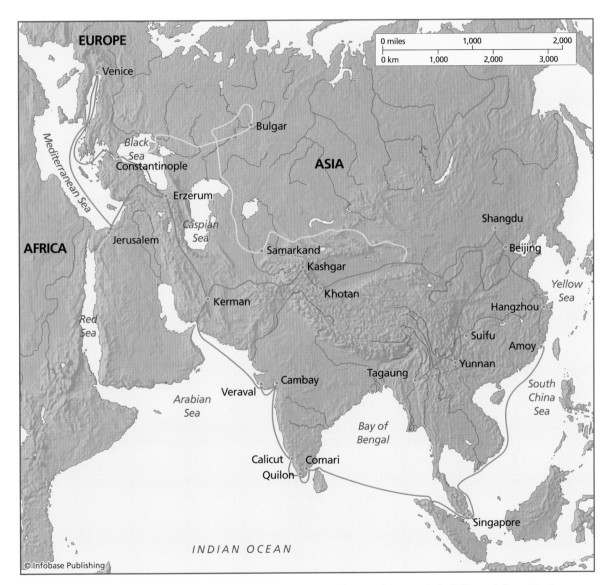

The route, shown by a heavy line, is that which Niccoló, Maffeo, and Marco Polo followed from Venice through the lands ruled by Kublai Khan and back to Venice. The pale line shows what is thought to be the route the two Polo brothers followed on their earlier journey.

The Polos arrived back in Venice in 1295 and became celebrities. In 1298 Marco was captured during a war between Venice and Genoa and was held by the Genoese as a prisoner of war. While he was imprisoned, Marco described his travels to a fellow prisoner Rustichello da Pisa, who was a writer of romances. They wrote *The Travels* together. The first edition was written in Old French and entitled *Le divisament dou monde* (The description of the world). That version is lost, but it had already been translated into Latin and from there into Italian. The book was soon translated into several other languages, and it was an immediate success.

The war between Venice and Genoa ended in 1299, and Marco was released. He returned to Venice, where his family had bought a large house out of the profits from their travels. Marco was now wealthy. In 1300 he married Donata Badoer, and they had three daughters, Fantina, Belela, and Moreta, all of whom later married into aristocratic families. Marco died at home in January 1324, leaving most of his possessions to be divided among his children.

MARCO POLO AT XANADU

Kublai Khan—the Great Khan as Marco Polo called him—was Mongolian and ruled at a time when the Mongolian Empire stretched from the River Danube on the edge of Europe to the shores of the China Sea, and from Lithuania in the north to the Himalayas in the south. He made his capital at what is now Beijing, but that was only his winter capital. In summer Beijing is very hot, and Kublai and his court of Mongolian nobles, who were not comfortable in the heat, escaped to his summer capital Shang-tu, to the north of the Great Wall. Tales of the glories of Shang-tu, also known as Xanadu, inspired the English poet Samuel Taylor Coleridge (1772–1834) to write *Kubla Khan,* which begins with the following lines:

> In Xanadu did Kubla Khan
> A stately pleasure-dome decree:
> Where Alph, the sacred river, ran
> Through caverns measureless to man
> Down to a sunless sea.
> So twice five miles of fertile ground
> With walls and towers were girdled round:

And there were gardens bright with sinuous rills,
Where blossomed many an incense-bearing tree;
And here were forests ancient as the hills,
Enfolding sunny spots of greenery.

Coleridge never visited Xanadu. Had he done so he would have seen that there are no sacred river, no caverns, and no sea—with or without sunshine. But Xanadu certainly existed and it was magnificent. Shang-tu is now in ruins, only rubble and pottery shards left to remind the visitor of its past grandeur. Kublai Khan founded the city in 1256, and it became known as the city of 108 temples. This is how it appeared to Marco Polo:

> In this city Kublai Khan built a huge palace of marble and other ornamental stones. Its halls and chambers are all gilded, and the whole building is marvellously embellished and richly adorned.

Where the palace abutted the city wall, another wall extended from the city wall away from the palace and enclosed a park with an area of about five square miles (13 km^2). The only access to the park was from the palace. The park contained springs and streams and there were lawns. Marco Polo's description continued as follows:

> Here the Great Khan keeps game animals of all sorts, such as hart, stag, and roebuck, to provide food for gerfalcons and other falcons which he has here in mew.* The gerfalcons alone amount to more than 200. Once a week he comes in person to insect them in the mew. Often, too, he enters the park with a leopard on the crupper of his horse; when he feels inclined, he lets it go and thus catches a hart or stag or roebuck to give to the gerfalcons that he keeps in mew. And this he does for recreation and sport.
>
> In the midst of this enclosed park, where there is a beautiful grove, the Great Khan has built another large palace, constructed entirely of canes, but with the interior all gilt and decorated with beasts and birds of very skilful workmanship. It is reared on gilt and varnished pillars, on each of which stands a dragon, entwining the pillar with his tail and supporting the roof on his outstretched limbs.

———————————————

*A mew is a cage for holding birds of prey.

As well as the game animals, there were leopards, lions, tigers, lynxes, wild asses, hippopotamus, and elephants in the park and eagles as well as gerfalcons (*Falco rusticolus*). According to Marco Polo, Kublai stayed at Shang-tu only during June, July, and August. The palace of canes was erected in time for his arrival and dismantled after his departure. Polo continued:

> When it comes to the 28th day of August, the Great Khan takes his leave of this city and of this palace. Every year he leaves on this precise day; and I will tell you why. The fact is that he has a stud of snow-white stallions and snow-white mares, without a speck of any other color. Their numbers are such that the mares alone amount to more than 10,000. The milk of these mares may not be drunk by anyone who is not of the imperial lineage, that is to say of the lineage of the Great Khan. . . . The astrologers and idolaters have told the Great Khan that he must make a libation of the milk of these mares every year on the 28th August, flinging it into the air and on the earth, so that the spirits may have their share to drink. They must have this, it is said, in order that they may guard all his possessions, men and women, beasts, birds, crops, and everything besides.

The ritual was to honor Kublai Khan's ancestors and it likely involved sacrificing one or more animals. Although today nothing remains of the park, archaeological excavations have revealed the layout of Shang-tu and its palace. There is no reason to suppose the park did not exist very much as Marco Polo described it.

QUEEN HATSHEPSUT AND THE FIRST EXPEDITION TO COLLECT ANIMALS

In displaying wild animals in cages and running free in a park, Kublai Khan was continuing a custom that was already extremely ancient. Egyptian pharaohs had owned zoos at least since 2500 B.C.E. Descriptions of them survive on the walls of tombs, on clay tablets, and in papyri—*papyrus* is a type of paper made from the papyrus plant (*Cyperus papyrus*), which is a sedge that grows 5–9 feet (1.5–2.7 m) tall; the Egyptians were using it by about 3000 B.C.E. The pharaohs stocked their zoos with such animals as hyenas, mongooses, monkeys, and antelope, and they made sure that the animals were well looked after. The Egyptians were the first people to keep newly arrived wild animals under controlled conditions while they acclimatized to their new lives.

The pharaohs enjoyed their zoos. They used them to demonstrate their wealth and prestige to visitors, and their scholarly priests learned much from studying the captive animals. They had little control over their stocking, however. The Egyptians had to rely on gifts from visiting delegations, which meant they received supplies of animals, but could not stipulate the species their collections lacked. The only solution to that difficulty was to send out expeditions for the express purpose of collecting wild animals according to a "shopping list."

In about 1480 B.C.E. Queen Hatshepsut, who ruled from 1479–1458 B.C.E., sent out the world's first expedition to collect specimens of plants and animals. The expedition was led by a general and consisted of five sailing ships, each of them 70 feet (21 m) long and carrying 210 men including 30 rowers, as well as Senenmet, the queen's steward. They sailed from Qusseir on the Red Sea coast for the land of Punt. Punt no longer exists, but historians believe it was in southern Sudan or Eritrea.

It was primarily a trading mission. The Egyptians needed frankincense, myrrh, and fragrant woods for making incense that they used in religious ceremonies and also for making cosmetics. They carried home the roots of frankincense (*Boswellia sacra*), preserving them carefully in baskets. The roots were planted in the grounds of the queen's mortuary temple at Thebes (modern Luxor). This is the first recorded instance of the deliberate introduction of imported plants.

The story of the expedition is told in carvings on the wall of the mortuary temple. A series of bas-reliefs shows the voyage and the arrival in Punt, where the Egyptians are received by the ruler of Punt and his wife—who has a pronounced curvature of the spine. As well as the goods they were seeking, the Egyptians returned with ebony, ivory, animal skins to be worn by Egyptian priests, and live animals. They also brought back Puntite adults and children.

Wealthy Egyptians were already familiar with some of the animals to be found in Punt, and Egypt had been trading with Punt for many years, but this was the first occasion on which they had gone to Punt to obtain live animals. They brought back dogs, monkeys, leopards, and giraffes. On their return the animals were kept in the queen's "garden of acclimation" at Thebes.

Hatshepsut was the daughter of Tuthmose I and Queen 'Ahmose. Following Tuthmose's death, his son became Tuthmose II, but Hatshepsut, his half-sister and wife, was really in control. Tuthmose II died after three or four years and was succeeded by Tuthmose III, who was Hatshepsut's stepson and married to her daughter, Neferure. He was a child so Hatshepsut ruled as regent. In fact, she took full control and ruled as pharaoh. It was rare for a woman to rule, though not unknown, and many images of Hatshepsut portray her wearing male dress and a false beard, probably to help establish her as a real king. She was a wise ruler who surrounded herself with reliable advisers, and she initiated many major building projects.

After Hatshepsut's death, Tuthmose III maintained a garden at Karnak, stocked with deer, domesticated livestock, birds, and exotic animals and plants brought from Syria. Akhenaten (ca. 1379–1361 B.C.E.) had a zoo with lions, antelopes, cattle, and a pond probably with fish and aquatic birds. The lions were kept in buildings, the antelopes in pens, and the cattle in an enclosure.

SOLOMON'S MONKEYS AND NEBUCHADREZZAR'S LIONS

According to the biblical account, in about 1000 B.C.E. King Solomon maintained herds of cattle, sheep, deer, and horses, as well as flocks of birds. David, Solomon's father, had formed an alliance with King Hiram (or Huram) of Tyre that continued into Solomon's reign, and Solomon used that trading agreement to obtain apes and peacocks from Tarshish (Spain). The second book of Chronicles records that:

> For the king's ship went to Tarshish with the servants of Huram: every three years once came the ships of Tarshish bringing gold, and silver, ivory, and apes, and peacocks. (2 Chronicles, 9:21)

Solomon was interested in agriculture and in animals. Little is known about his life apart from what is written in the Old Testament. The first book of Kings states that "Solomon had 40,000 stalls of horses for his chariots, and 12,000 horsemen," as well as straw for the horses and dromedaries (although in the second book of Chronicles the number of stalls for his horses is reported as 4,000).

Nebuchadrezzar II ruled Babylonia from 605 to 562 B.C.E. His name, in the Akkadian language, means "The god Nabu has guarded the estate." He was the son of Nabopolassar, who founded the Chaldean dynasty, and a few months before his father's death, Nebuchadrezzar defeated the Egyptians, thereby extending Babylonian rule to all of Syria and Palestine. He married Amytis, daughter of the king of Media, a country in what is now northern Iran, and in her honor he built a palace at the top of a series of stepped terraces, situated between the double walls of Babylon. There were gardens on all the terraces, and underground canals supplied water for domestic use and irrigation and drainage. These royal gardens came to be known as the Hanging Gardens of Babylon.

Outside the garden terraces there was more extensive parkland, and that is where the king's wild animals were kept. These included lions, which were important symbols. Solomon's throne, made of ivory covered in gold, had six steps leading up to it, with two statues of lions on each step. Lions were widely represented on palace furniture.

Years later, in 539 B.C.E., Nebuchadrezzar's grandson Belshazzar was killed by Darius, king of Media, and the Babylonian throne usurped. That is when courtiers concocted charges against Daniel and demanded that Darius have him cast into a den of lions, according to the law. The Old Testament book of Daniel, chapter 6, describes how this was done, and how Daniel survived. Events of this kind very likely happened, because rulers had been keeping lions in parklands for centuries. Lions roamed the parklands of the Assyrian king Tigathpileser I (reigned 1116–1076 B.C.E.). Kings hunted them to demonstrate their strength and courage.

The lions were Asiatic lions (*Panthera leo persica*), a subspecies that were once found in the Balkans and all across the Middle East. Today they survive only in the Gir Forest in India. Samson, in the Old Testament book of Judges, encountered an Asiatic lion while on his way to ask for the hand in marriage of the woman with whom he had fallen in love. Samson tore the lion apart with his bare hands (Judges 14:6).

GREEK BIRD GARDENS

On his long journey home following the fall of Troy, the Greek hero Odysseus was imprisoned for seven years by the nymph Calypso on her island of Ogygia, which is thought to be Gozo, the island adjacent to Malta in the Mediterranean. The god Zeus sent his son Hermes, the messenger, to call on Calypso and tell her that she had held Odysseus for long enough and it was time to free him. Book V of Homer's *Odyssey* (in the translation by E. V. Rieu) described his arrival in the following words:

> So Hermes rode the unending waves, till at length he reached the remote island of Ogygia, where he stepped onto the shore from the blue waters of the sea and walked along till he reached the great cavern where the Nymph was living. He found the lady of the lovely locks at home. A big fire was blazing on the hearth and the scent from burning logs of split juniper and cedar was wafted far across the island. Inside, Calypso was singing in a beautiful voice as she wove at the loom and moved her golden shuttle to and fro. The cave was sheltered by a verdant copse of alders, aspens, and fragrant cypresses, which was the roosting-place of feathered creatures, horned owls and falcons and garrulous choughs, birds of the coast, whose daily business takes them down to the sea.

To Greek ears, this description would have suggested a romantic, strange land. Their knowledge of extensive ornamental gardens and parklands came from the east, in tales brought by travelers returning from Asia, who had seen the gardens of such fabled cities as Babylon. Back home, the Greeks lived in city-states, and their cities were enclosed by defensive walls. Inside the walls, the buildings were densely crowded, and there was little room for gardens. It is not that there were no plants, but plants grown in the city had practical uses. There were vineyards and fruit orchards, vegetable plots and, of course, there were olive groves to provide the olives and oil that were an essential part of the Greek diet. Flowers were grown, but mainly in containers.

Many Greek citizens kept various species of birds as pets. Aviaries occupy less space than gardens. They also kept pet monkeys, hedgehogs, and nonvenomous snakes, and they had dogs of course. There is no evidence that they kept domestic cats. The Egyptians cherished cats. Indeed, they valued them so highly that for at least 1,000 years after the cat was domesticated it was illegal to carry one out of Egypt. Some time after 460 B.C.E., the Greek historian Herodotus visited

Egypt and was fascinated by the cats he saw there and the role cats played in Egyptian life. It seems unlikely that he would have taken so much interest if cats were as familiar to him as they are to everyone in the modern world. Rodents stole and contaminated food in Greek stores, and the Greeks had mousers and ratters to help control vermin, but they used ferrets, possibly polecats, which are domesticated ferrets, rather than cats. Interestingly, the words for people who love or hate cats are *ailurophile* and *ailurophobe,* respectively, and the Greek word *ailouros* means ferret. The first record of a domestic cat in Greece is a marble bas-relief in the National Archaeological Museum of Athens. It shows a cat on a leash facing a dog, also on a leash, with two people watching who seem uncertain what will happen when a cat and dog meet.

Many animals had religious significance, and both wild and domesticated species were held in the grounds of temples. The animals would have required food and shelter, so a garden would have been needed to grow the plants. Many temples had bears and lions, both of which lived wild in Greece. A few temples had tigers, imported from much farther afield. At religious festivals the temple animals were paraded through the streets. Those that were tame were led on leashes, and those that could not be tamed were drawn along in cages.

In parallel to the parades of temple animals, there were also showmen who specialized in wild animals as entertainment. Professional trainers prepared some animals for performances, and other animals were simply exhibited to terrify onlookers. Philosophers studying animals relied on temple animals for their specimens, although Alexander the Great arranged for animals captured by his soldiers to be sent back to Athens for Aristotle to study at his Lyceum (see "Aristotle and the Importance of Observation" on pages 18–21). The tradition of sending animals captured abroad back to Greek cities continued after Alexander's death.

Although she ruled Egypt and styled herself pharaoh, Cleopatra VII Philopater (51–30 B.C.E.) spoke Greek more often than Egyptian. According to some reports, Cleopatra was the last Egyptian ruler to bother learning the Egyptian language. Later kings employed interpreters and translators. Kings of Macedonian descent had ruled Egypt since the country was conquered by Alexander in 332 B.C.E., and in 305 B.C.E. the Macedonian general Ptolemy I Soter (ca. 367–282 B.C.E.) pronounced himself pharaoh. From that point on, Egypt remained a Ptolemaic kingdom until the death of Cleopatra and the Roman invasion and occupation that followed.

From 285 B.C.E. until 246 B.C.E., the pharaoh was Ptolemy II Philadelphus (308–246 B.C.E.), the son of Ptolemy I Soter. For the first three years he ruled in partnership with his father. During his reign Egypt was engaged in several wars, but Ptolemy II is not regarded as a particularly belligerent king. He was highly intelligent, sponsoring scientific research and expanding the large library at Alexandria, the capital of Ptolemaic Egypt.

Ptolemy II maintained a magnificent court and gave displays of his immense wealth. He also supported a large collection of animals. These were used in religious processions, according to Greek custom. To celebrate his coronation, Ptolemy arranged a procession through Alexandria. In book V of his *History,* the Greek writer Athenaeus, who lived in about 200 C.E. and long after the event, described the procession. First there were groups of men honoring several gods, culminating in the procession for Dionysus. Athenaeus continued as follows:

> The next was a four-wheeled wagon 14 cubits high and eight cubits wide; it was drawn by 180 men. On it was an image of Dionysus—10 cubits high. He was pouring libations from a golden goblet, and had a purple tunic reaching to his feet.... After many other wagons came one 25 cubits long, and 15 broad; and this was drawn by 600 men. On this wagon was a sack, holding 3,000 measures of wine, and consisting of leopards' skins sewn together. This sack allowed its liquor to escape, and it gradually flowed over the whole road.... An endless array of similar wonders followed; also a vast number of palace servants displaying the golden vessels of the king; 24 chariots drawn by four elephants each, the royal menagerie—12 chariots drawn by antelopes; 15 by buffaloes; eight by pairs of ostriches; eight by zebras; also many mules, camels, etc., and 24 lions.

A cubit was the length from the elbow to the fingertips, about 18 inches (45.7 cm). So the first wagon mentioned in the quotation above was 21 feet (6.4 m) high and 12 feet (3.7 m) wide, and the figure of Dionysus was 15 feet (4.6 m) tall. The huge wagon, drawn by 600 men, was 37.5 feet (11.4 m) long and 22.5 feet (6.9 m) wide. Other accounts of Ptolemy's procession list 2,000 bulls with horns painted or covered with gold, 200 Arabian sheep, 16 cheetahs, seven pairs of oryxes, 23,200 horses, one bear, one giraffe, cages of parrots and peacocks, and 150 men carrying trees with birds perched in them.

ROMAN PETS, MENAGERIES, AND CIRCUSES

Ptolemy's parade of animals may seem extreme, but it was modest compared with the Roman parades and entertainments that used animals imported from all parts of the empire, especially from Africa. Elephants were popular, and they played an important part in Julius Caesar's (100–44 B.C.E.) triumph—victory parade—in 46 B.C.E. From 49 B.C.E. until 45 B.C.E., Roman armies were fighting a civil war that began on January 10–11, 49, when Julius Caesar, then a general, invaded Italy by leading his troops across the River Rubicon, which marked the border with Gaul. The general leading the opposing army was Pompey the Great (106–48 B.C.E.), widely recognized as the greatest of all Roman generals until his defeat by Caesar in 48 B.C.E. Pompey was assassinated later that same year. The victorious Julius Caesar celebrated his victory with a triumph through the streets of Rome and up the steps of the Capitol. It included 40 tame elephants, each carrying a blazing torch in its trunk, and the elephants climbed the steps as well as the soldiers and their general. Elephants symbolized Roman imperial power, and the torches were to shine brightly on the general's victories.

Elephants were also used in warfare, most famously by the Carthaginian general Hannibal (247–183 or 182 B.C.E.). In 218 B.C.E. Hannibal led his army across the Alps in an attack on Rome. His force included 38 elephants, but few of them survived the crossing. Carthaginian coins of the time depict the elephants, showing they were African elephants (*Loxodonta africana*), although some historians think they might have been Indian elephants (*Elephas maximus*), some of which had been imported into North Africa.

Mainly, though, the Romans used elephants for entertainment. They were dressed in ludicrous costumes, trained to walk tightropes, throw weapons, perform cumbersome dances, and were even made to sit at a dinner table beside humans. Bears were also dressed in costumes and made to perform. Animal performances took place either in the street or in arenas such as the Circus Maximus, which could seat an audience of 150,000. The Circus Maximus, the original circus, is where chariot races were held—it still exists and is an important tourist attraction in Rome. The Circus Maximus was also the venue for staged hunts, in which wild animals were pursued and killed with spears and knives, and for fights to the death between animals. Elephants sometimes fought bulls or other elephants, and there were fights involving lions, bears, rhinoceroses, and other beasts.

Circuses—events held in the Circus Maximus—were later transferred to the Colosseum, which opened in 80 C.E., and they continued to be called circuses. The Colosseum is also where gladiatorial combats were staged before up to 50,000 spectators. Gladiators fought each other or fought against wild animals that were kept in cages beneath the stadium. A contest with a wild, half-starved animal was also a method of public execution. The games held in the Colosseum were hugely popular. The spectacle that opened the Colosseum lasted for 100 days, and at its end more than 9,000 animals and 10,000 condemned prisoners had been killed. By the second century there were more than two of these public entertainments every week. Not all were brutal. They included plays and vaudeville as well as combat.

The Roman public had an insatiable appetite for such events, but the well-to-do also kept pets and small menageries. Dogs were popular, as were ponds stocked with colorful fish, aviaries, and caged animals such as gazelles. Some people even kept pet lions, usually with their claws and teeth removed.

Bears were captured in Italy, and many of the other animals were imported from Africa. Transported in crates and often suffering injuries inflicted during their capture, they formed part of a highly lucrative trade. There was a good market among Romans for anything exotic from the farther reaches of the empire, and marketable products included animal parts as well as the live animals themselves.

THE MENAGERIE OF OCTAVIUS AUGUSTUS

The Roman emperor Augustus (63 B.C.E.–14 C.E.), for whom the month of August is named (it was previously called Sextilis, the sixth month in the Roman calendar), continued the tradition of keeping a menagerie. Augustus was a highly intelligent, cultured man, who took a keen interest in the acquisition of knowledge. So, although his animal collection was a menagerie rather than a zoo in the modern sense (see the sidebar below), the emperor probably studied its occupants and learned something from them. It was a large collection, reputedly containing 420 tigers, 260 lions, cheetahs, leopards, 36 crocodiles, one rhinoceros, one hippopotamus, as well as elephants, bears, seals, eagles, and many others.

Emperors needed exotic animals for parades and ceremonies and for official celebrations that involved public displays of hunting and fighting. Augustus was no exception, and over the 58 years of his

THE DIFFERENCE BETWEEN A MENAGERIE AND A ZOO

From the very earliest times rulers and wealthy aristocrats have assembled collections of wild or exotic animals and maintained them in parks or gardens adjacent to their palaces. The animals were difficult and costly to acquire and expensive to house, feed, and care for. Consequently, the animal collection was an obvious demonstration of the wealth and power of its owner. The animals were not kept in captivity in order for natural historians to study them, but for entertainment, and the animals themselves were usually chosen because, though exotic, they were not so violent as to be difficult or dangerous to manage.

A collection of such animals that is maintained for entertainment or to impress visitors is a *menagerie*. During the Middle Ages, most European monarchs and many nobles owned menageries. In the 18th century, traveling collections of animals were popular in Europe and in America, and these were also known as menageries. They were a form of popular entertainment. In 1716 in Boston a lion became the first wild animal to be exhibited in America. Other animals quickly followed. In America these exhibitions were combined with circus shows based

on clowns, acrobats, and other human as well as animal performers. A single ticket allowed entrance to both the circus and the menagerie, and the menagerie was as popular and important as the circus.

In 1826 the Zoological Society of London was established to promote the scientific study of animals—*zoology*—and animal physiology. In 1828 the society installed its collection of animals in Regent's Park, London, and opened what it called its *zoological gardens* to members of the society and their guests. In 1847 the society opened the gardens to the general public, who paid an entrance fee that was used to support the maintenance of the animals and grounds and scientific research. It was also in about 1847 that zoological garden was first abbreviated to *zoo* in a popular song.

The London zoo was the first to be called a zoo, but it was not the first to be established. That was probably the Ménagerie du Jardin des Plantes, which opened in 1793 in Paris as part of the Muséum d'Histoire Naturelle (Natural History Museum). It was laid out as a park, but its animals were in cages—as they were in London.

rule, from 44 B.C.E until his death in 14 C.E., more than 3,500 of his menagerie animals lost their lives in 26 of these celebrations.

Augustus was born in Rome into a wealthy family, with the name Gaius Octavius. His mother, Atia, was the daughter of Julia, Julius Caesar's sister, so Octavius was Caesar's nephew. Caesar adopted him as his son, encouraged him, advanced his career, and nominated him as his successor and heir to his wealth. When Caesar was murdered in 44 B.C.E., Octavius was 19 years old and a student in Illyria—a region of the Balkans approximately corresponding to modern Albania, Croatia, Serbia, and Bosnia-Herzegovina. On receiving the news of Caesar's death, he immediately returned to Rome and changed his name to Gaius Julius Caesar. Augustus was an additional honorific name given to the emperor in 27 B.C.E. by order of the senate. The name meant sacred or venerable—it has come into the English language as "august." The emperor did not use it of himself. All succeeding emperors also bore this name. In 42 B.C.E., he deified Julius Caesar and henceforth styled himself "son of the divine" (*divi filius*) and insisted on being addressed as Caesar. Caesar, which was Julius Caesar's surname, was adopted as a title by all subsequent Roman emperors and came to mean "emperor" in several European languages (e.g., German *kaiser,* Russian *czar*).

It was not easy for this young man to seize and retain control over the disparate factions that were jockeying for power in the years following Caesar's political assassination, and he was ruthless and often brutal in destroying those who opposed him. After the assassination, Marc Antony (Marcus Antonius, ca. 82–30 B.C.E.) had seized power, and he resented the arrival of Octavius, but because Caesar had personally nominated him, Octavius retained the loyalty of the army and Antony was unable to resist. The two reached a compromise, but Octavius was always the more skillful negotiator and politician. Marc Antony married Cleopatra (see "Greek Bird Gardens" on pages 39–43) and after her suicide he also took his own life.

Once his rule was secure, Octavius's personality changed, and he became truly statesmanlike. He had a sense of humor and tolerated criticism. Augustus was particularly skillful in reading and responding to public opinion. He enlarged the empire, but also brought peace throughout it, known as the *pax Romana.* Although he was "son of the divine," Augustus did not permit himself to be worshipped, and, despite being the most powerful ruler in the world, the house he lived in was large but not a palace. By the time of his death, Augustus was widely respected and greatly loved. He fell ill during a journey to the island of Capri and died at Nola, in mainland Italy, on August 19, 14 C.E.

NERO AND THE GROUNDS OF HIS GOLDEN HOUSE

Augustus was succeeded as Roman emperor by Tiberius (42 B.C.E.–37 C.E.), followed by Caligula (12–41 C.E.), and Claudius (10 B.C.E.–54 C.E.). In the year 54, Nero Claudius Caesar (37–68 C.E.), at the age of 16, became the emperor. Each of these emperors sought to live in the style of Augustus. All of them kept menageries and all of them encouraged games that grew increasingly brutal and profligate. During one of the games held to honor Caligula, 400 bears were killed on a single day.

Nero began his reign with a promise to end secret trials and government corruption and to give more power to the senate. He largely kept his word. Nero also hated signing death sentences. On the other hand, it was not long before he was devoting himself almost entirely to pleasure. He thought of himself as a champion charioteer, a talented poet, and an accomplished performer on the lyre. He gave public performances and appeared in plays. But as the years passed, the emperor became increasingly despotic and paranoid, and, as opposition to him increased, he rid himself of his enemies in what became a reign of terror.

In 64 C.E. a major fire destroyed two-thirds of Rome. Nero did what he could to organize the rescue of those affected and to find accommodation for the many people rendered homeless, but by this time his vanity and indecorous behavior had made him very unpopular. Many people believed he had started the fire himself, although there is no evidence to support that. Nero responded by shifting the blame onto the Christians, thereby triggering the intermittent persecution of Christians that followed.

Since he had first come to power, Nero had cherished grandiose dreams of rebuilding Rome. The fire gave him an opportunity, and he commandeered the area destroyed by the flames to construct a park and within it a Golden House—*Domus Aurea* in Latin. The Golden House was not a palace in the conventional sense, but a place for holding parties for which Nero was famous. It had 300 rooms, but there were no bedrooms and archaeological researches have so far discovered no kitchens and no toilets. Parts of the main building were covered in gold leaf—hence the name—and rare seashells, ivory, and gemstones decorated the interior ceilings. Outside the entrance there was a statue of Nero, possibly in the guise of the sun god, approximately 100–115 feet (30–35 m) tall. The statue stood at one end of a covered arcade about 4,900 feet (1.5 km) long that linked the golden house to the private villa where Nero stayed when he was in the park. Nero's palace remained on the Quirinal Hill, one of the seven hills on which Rome is built; the Quirinal Palace is still the official residence of the Italian president.

The entire estate covered three of Rome's seven hills (the Palatine, Esquiline, and Caelian hills). Its park aimed to reproduce the countryside in the heart of the city—it was fake countryside. It had an artificial lake large enough for ships to sail on, with fake coastal villages along its shores. There were vineyards, areas of woodland, and pastures. Domestic and wild animals roamed freely. The poet Martial (Marcus Valerius Martialis [ca. 41–ca. 102 C.E.]) commented that: "One house took up the whole city of Rome."

The stock of animals supplied the needs of the games, of course, and Nero's games were excessive. On one occasion he had 400 tigers fighting against elephants and bulls.

HENRY I AND THE FIRST ENGLISH MENAGERIE

In the year 1066 Anglo-Saxon rule in England ended with the Norman Conquest that brought William I (1027–87), known as the Conqueror, to the throne. William had nine children. He was succeeded by his third son, who became William II, also known as William Rufus because of his florid complexion (ca. 1056–1100). When William II died, the Conqueror's fourth son became Henry I (ca. 1069–1135). In the year he became king, Henry married Edith of Scotland (who changed her name to Matilda), the daughter of Malcolm III of Scotland and great-niece of Edward the Confessor, who had been an Anglo-Saxon king. The marriage united the Norman and Anglo-Saxon royal families.

In 1106 Henry defeated Robert Curthose (ca. 1051–1134), the Conqueror's eldest son, to become Duke of Normandy. Robert had been very unpopular and the people rose against him, appealing to Henry to come to their aid. When the king had united England and Normandy, he imprisoned Robert, who remained in captivity for the rest of his life.

In April 1105, during his war with Robert Curthose, Henry captured the Norman town of Caen and marked the victory with a parade through the streets. His parade included camels, an ostrich, a young lion, and a lynx. The people lining the street to watch the parade were delighted, but the aim was not to entertain: Henry was demonstrating his power. A king who could control such fierce beasts would have no trouble controlling unruly subjects.

Henry was a skilled diplomat and highly efficient ruler. He was also well educated, being the first English king since Alfred the Great (ca. 849–ca. 899) who could read and write, in Latin as well as English. London was the capital of England, but kings have always had palaces in various parts of the country. Henry had a palace at Woodstock, near Oxford. It was a peaceful place where he went to relax and escape from the affairs of state, but without moving too far from London. The map on page 50 shows the location of Woodstock in relation to London, Oxford, and other cities. (Blenheim Palace, close to Woodstock, is where Winston Churchill was born.) Originally the palace was probably a hunting lodge inside a royal forest. Both William the Conqueror and William Rufus stayed there while hunting in the forest, but Henry converted it into something more appropriate for a royal residence. Mary I imprisoned Princess Elizabeth (later Elizabeth I) there in 1554. The palace remained in use until the 17th century.

Henry had a stonewall seven miles (11 km) long built to enclose part of the land surrounding the palace, and, in about 1125, he began to stock the resulting enclosure with imported animals. These included lions, leopards, lynx, camels, owls, porcupines, and an ostrich. This was the first menagerie in England, and Henry took a keen interest in the animals. No doubt he used them to impress visitors, but he also took a keen interest in them. The porcupine, given to him by William VI of Montpellier (ca. 1100–ca. 1162), was a particular favorite. There are records showing that Henry de la Wade of Stanton Harcourt was required to spread feed for the king's beasts and to mow a meadow in the park at Woodstock and carry the hay from the meadow to feed the animals. Usually hay was supplied from Oxford, but in 1201 floods ruined the crop and the Woodstock estate had to provide it.

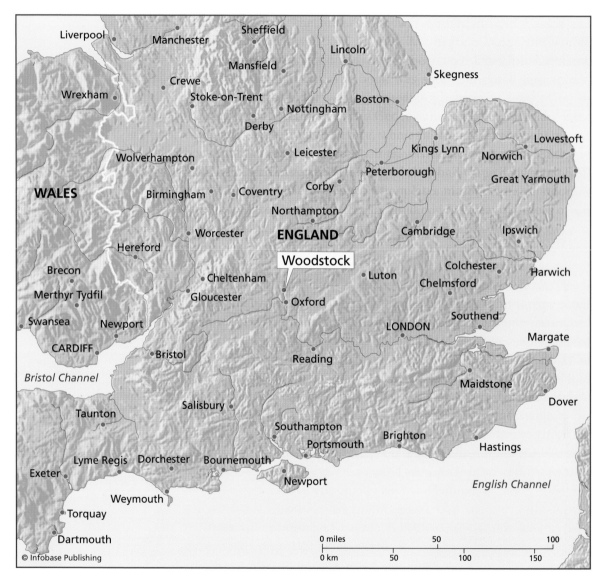

Woodstock is a town with about 3,000 inhabitants located 73 miles (117 km) from London and 8 miles (13 km) from Oxford. The name is Saxon and means "clearing in the wood." In the 11th century the area was a royal forest. Henry I is believed to have kept England's first menagerie in a park on the southern side of the modern town.

KING JOHN AND HIS MENAGERIE
AT THE TOWER OF LONDON

In about 1204, King John (1166–1216) established a menagerie at the Tower of London. According to Geoffrey Parnell, the present keeper of the tower, John is known to have had animals on his lands in Normandy and when he lost Normandy to King Philippe-Auguste of France three shiploads of animals were transported to England, most probably to be taken to the Tower. The collection would have consisted of gifts from other rulers. There would have been lions—always popular as symbols of royal power—as well as tigers, elephants, bears, porcupines, and other exotic animals. It is likely that John also moved the animals he inherited in the Woodstock menagerie established by Henry I to the Tower.

The Tower of London is officially the palace and fortress of the British sovereign, although the last monarch to live there was James VI of Scotland and I of England (1566–1625). The central part of the complex is the White Tower, shown in the illustration on page 52. Work on fortifications at the site began around Christmas Day, 1066, immediately after William the Conqueror had seized power, and construction of the White Tower began in 1078. Over its long history, the White Tower has housed an armory, the royal treasury, the royal mint, and the public records office. The English (but not Scottish) crown jewels are still held there. It has also been a prison and place of execution, mainly for high-status criminals. The Tower is located on the northern bank of the River Thames, and today it is a popular tourist attraction.

John was the most unpopular ruler England has ever had and is perhaps best known as the treacherous, vindictive enemy of Robin Hood. The youngest—and favorite—son of Henry II, John came to the throne on April 6, 1199, following the death of his brother Richard I (1157–1199). He quarreled with Pope Innocent III over the appointment of the Archbishop of Canterbury and imposed a tax on his barons in place of their traditional feudal duty to provide knights to fight for him. That caused the barons to rebel, forcing him to sign the Great Charter (Magna Carta) in 1215—although since John signed under duress the pope allowed his appeal against the terms of the charter, absolving him of any need to observe them. By the end of his reign, England was on the verge of civil war. John was a highly competent administrator, however, always well informed, and he

A popular tourist attraction, the Tower of London has been a palace and, more famously, a prison. Construction of the fortifications began in 1066, and work on the central keep, known as the White Tower and shown here, began in 1078. The White Tower is at the center of a concentric series of fortifications on the northern bank of the River Thames and close to the city wall built by the Romans. *(Taxi)*

was open-minded. He took a keen interest in the law and sometimes sat as judge in court proceedings, applying the law fairly. His quarrel with the barons led them to invite Prince Louis of France to take the English throne, and the French invaded. John opposed them, but he was no warrior, and, while retreating through eastern England, avoiding areas held by the rebels, the convoy carrying his baggage crossed a low-lying coastal area and was lost when the tide rose faster than expected. The baggage included the crown jewels and their loss badly affected the king's physical and mental health. He died from dysentery at Newark Castle, on the border between the counties of Lincolnshire and Nottinghamshire, on October 18 or 19, 1216, and was buried in Worcester Cathedral.

In 1235, John's son and successor, Henry III (1207–72), received a wedding gift of three leopards (or possibly the lions depicted on his coat of arms—and now the royal coat of arms) from Frederick II (1194–1250), the Holy Roman Emperor. In 1264, these animals were transferred to the Bulwark, near the principal western entrance to the Tower, and the semicircular Bulwark was renamed the Lion Tower. Archaeologists found the skull of a lion in 1936–37 during excavations of the site of the moat, close to the Lion Tower. The skull was moved to the Natural History Museum, where it remains, and it was dated as having lived between 1420 and 1480. The Lion Tower was demolished in the 19th century, although Lion Gate remains and the site of the tower is marked. During the reign of Elizabeth I (1533–1603; reigned 1558–1603), the menagerie was occasionally opened to the public and by the 18th century it was open all the time. Visitors paid an entrance fee of three halfpence (1.5 pennies in pre-decimal money), and those unable to pay cash could provide a cat or dog to feed the big cats. It was at the Tower menagerie that the English poet William Blake (1757–1827) saw the tiger that inspired his poem *The Tyger*, which he wrote in about 1794. It begins with the following lines:

> Tyger Tyger, burning bright,
> In the forests of the night:
> What immortal hand or eye,
> Could frame thy fearful symmetry?

The animals in the Tower Menagerie were not treated to a standard that would be thought acceptable today. Archaeological excavations have exposed the sites of lion cages 6.5 × 10 feet (2 × 3 m) in size. An adult lion can reach a body length of 8 feet (2.5 m), so their accommodation was cramped. Records from the 15th century show that at that time they were fed on sheep. Cheetahs were fed on boiled meat and grain meal in summer and on gruel and hot mashes of cereals in winter. Ostriches, in contrast, were fed a diet rich in nails, because they were believed to have a huge appetite for iron. In 1623, James I received a gift of an elephant. It was fed on nothing but one British gallon (1.2 U.S. gallon; 4.5 liters) of wine a day. Bones of dogs have also been found at the Tower. Dogs were used to bait larger animals, including lions and tigers, as a form of entertainment.

London Zoo, located in Regent's Park, was first stocked with animals from the menagerie at the Tower of London. The zoo now has one of the largest animal collections of any zoo and is very active in animal conservation. This is Raja, the zoo's male Komodo dragon. (*Varanus komodoensis*); he is seven feet (2.2 m) long. *(2004 Getty Images)*

Among the many species in the menagerie, there were also monkeys, hyenas, a jaguar, wolves, mongooses that proved very useful in disposing of the rats infesting the buildings, antelope, birds, and several snakes. Alfred Cops, a new director who arrived at the menagerie in 1822, found its stock depleted, however, and the establishment in a poor state. He introduced new stock, but the accommodation was unsuitable for the menagerie animals and their upkeep was proving expensive. When the Zoological Society of London was established in 1826 with the aim of assembling a collection of animals, the authorities decided to transfer the stock from the Tower menagerie to the society. The entire stock was transferred in 1831 to London Zoo, which had opened in 1828, with the proviso that any unwanted specimens be sent to Dublin Zoo, which had opened in 1830. Today, London Zoo holds one of the world's largest animal collections. The illustration above shows Raja, a male Komodo dragon (*Varanus komodoensis*), which the London Zoo acquired in 2004.

THE ROYAL MENAGERIE AT VERSAILLES

As the European powers expanded their empires, European monarchs were able to stock their animal collections from all corners of the world. Royal menageries were popular with royalty, but when the French king established one at the largest and most splendid of his palaces, the menagerie and its occupants became fashionable. In the 1780s, male followers of fashion were wearing suits with zebra stripes.

It all began in the 17th century. At that time Versailles was a village 12 miles (20 km) southwest of Paris. Louis XIV (1638–1715), known as the Sun King (*le roi soleil*), enjoyed hunting in the royal forest around Versailles, staying at a hunting lodge that his father had built there in 1624. Louis decided it would be an appropriate

The palace of Versailles was built in 1624 as a hunting lodge and expanded into a palace in 1669. It was the official residence of French kings from 1682 until 1790. As well as the formal gardens, its grounds held the royal menagerie, which was a precursor of modern zoological gardens. *(Rosine Mazin/Photo Researchers)*

place to relocate his entire court. First, though, he needed a palace, and in 1669 he commissioned the architect Louis Le Vau (1612–70) to design the buildings, Charles le Brun (1619–90) to design the interiors, and André Le Nôtre (1613–1700) to design the surrounding landscape and gardens. The court began moving to the new palace at Versailles in 1678. The move was complete and the court officially installed there on May 6, 1682. The illustration on page 55 shows the palace as it is today.

The palace was certainly large, dwarfing Nero's Golden House (see "Nero and the Grounds of His Golden House" on pages 47–48). It had 700 rooms with 67 staircases and a total floor area of 550,000 square feet (51,210 m^2). The roof area covered 27 acres (11 ha), and there were more than 2,000 windows.

There was also a menagerie. Work on that began in 1663, and the first animals arrived the following year. The grounds around the Versailles palace contained a number of pleasure houses—buildings used exclusively for entertainment—and the menagerie centered on the first of these to be built, the Pavillon de la Lanterne (Lantern Pavilion). The Sun King's animals were to be primarily entertainers.

Visitors entered the building through a forecourt. Inside, a staircase led to an upper floor, with rooms to the left and right and a central passage leading to the eight-sided Lantern Pavilion, which contained a single room. The passage entered the pavilion at one of its sides. Windows in the remaining seven sides opened onto balconies overlooking an octagonal courtyard that surrounded the pavilion on seven sides. Each section of the courtyard held an animal enclosure. A staircase inside the pavilion led to a ground floor decorated with seashells to resemble a grotto, from where visitors could enter the center of the courtyard to view the animals at closer quarters. There were six marble columns in the courtyard, each with a fountain.

The animal enclosures were no less elaborate. They were planted with grass and contained ponds, fountains, and buildings to shelter the animals, iron railings and walls separated each enclosure from its neighbors, and classical architecture ornamented the rear walls. When it opened, the menagerie contained mainly exotic birds, but there was also accommodation for horses, gazelles, and other large mammals. Early in the 18th century Louis expanded the menagerie by adding another set of enclosures to house lions, tigers, leopards, and other species.

Life at the court of the Sun King was ruled by elaborate and inflexible rules of etiquette, so that excessively polite courtiers spent their days performing rituals that were almost theatrical. The king loved theatrical performances and would often hold them in the grounds. They usually began with a procession led by the king and queen, followed by displays in which actors and musicians portrayed themes from classical literature or the seasons of the year, assisted by animals from the menagerie.

Although entertainment was the principal purpose of the menagerie, as the years passed and the size of its collection grew, the animals became the subjects of much scientific research. At its peak, the Versailles menagerie housed 222 species. Louis XIV died in 1715, and Versailles became the property of Louis XV (1710–74), his successor. With the change of monarch, the fortunes of the menagerie began to wane. Louis XV had little interest in the animals and, because it was the king who dictated the fashion, the menagerie ceased to be fashionable. By the time Louis XVI (1754–93) came to the throne in 1774, the menagerie and its buildings were in a poor condition. As the animals went out of fashion, so the funds to care for them dwindled.

Following the French Revolution of 1789, there was popular resentment against the Versailles menagerie. Why should these animals live in luxury, revolutionaries asked, while people are starving? The palace of Versailles was in the hands of the revolutionaries, and in 1789 a crowd arrived at the entrance to the menagerie demanding that the imprisoned animals be set free. The director of the menagerie pointed out that if certain of the animals were to be released, their first action would likely be to devour their liberators. So the lions, tigers, and other large carnivores remained behind bars. Some animals were released, and others were killed. According to Bob Mulland and Garry Marvin, in their book *Zoo Culture,* this is the only recorded instance in history of a popular attempt to free zoo animals.

Some of the animals were left behind, and the Versailles estate manager arranged with Jacques-Henri Bernardin de Saint-Pierre (1737–1814), the writer and botanist who had been director of the Jardin des Plantes (botanical garden) since 1793, to have these transferred to his charge. Bernardin de Saint-Pierre had been urging the authorities to allow him to establish a national menagerie at the gardens, so he welcomed the news.

The botanical garden, covering 58 acres (23 ha) began in 1626 as the Jardin du Roi (royal garden), growing medicinal and culinary herbs and located beside the River Seine in the center of Paris. After the revolution its name was changed to the Jardin des Plantes, and in 1793 the National Assembly ordered the creation of a menagerie in the garden, decreeing that all exotic animals in private hands, including animals belonging to circuses, fairs, other traveling shows, and street performers as well as wealthy collectors, must be given to the national menagerie either alive or stuffed. By the middle of the 19th century, the national menagerie held the largest animal collection in Europe. It was open to the public from the start and entrance was free, but scientific research was its primary purpose. It had ceased to be a menagerie and had become the world's first zoo (see the sidebar "The Difference between a Menagerie and a Zoo" on page 45). The zoo is now part of the Muséum national d'Histoire naturelle (National Natural History Museum).

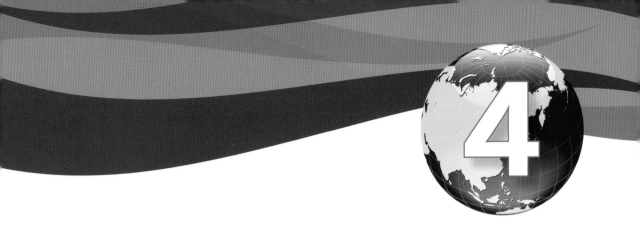

Travelers' Tales

Aesop was a slave who lived in Greece from 620 B.C.E to 560 B.C.E. He told very brief and often amusing stories about animals and sometimes about plants that have been translated into most languages. Over the centuries Aesop's fables have delighted many generations of children and inspired many authors.

Each of his tales carries a message. For example, an ass and a lion formed an alliance to capture other animals as prey. The ass gave the lion the benefit of his greater speed and the lion contributed his strength. After they had caught a number of animals, the lion divided the spoils into three shares. The lion said he would take the first share, because he was king of the beasts. He would also take the second share, due to him as a partner in the hunt. The third share, he explained, would become a source of great evil to the ass unless the ass relinquished any claim to it and made off as fast as he could. The moral? Might makes right. Another fable tells of some pigeons that were terrified when a kite appeared and called on the hawk to protect them. The hawk agreed, but as soon as the pigeons admitted him into their dovecote the hawk killed more of them in one day than the kite would have killed in a year. The moral? Beware that the remedy is not worse than the disease. One of the most famous of Aesop's fables concerns a crow that had stolen a piece of meat and taken it to her perch high in a tree when a fox determined to take it from her. The fox flattered the crow, expressing admiration for her shape and plumage, but lamenting that if only her voice could match her appearance she would undoubtedly be regarded as the queen of birds. Determined to show that her voice was the equal of any bird, the crow opened her bill and let out a loud cry. This made her drop the meat, which the fox seized. The fox told the crow that her voice was good enough, but she lacked wits.

There are hundreds of these fables in which animals and plants possess certain human characteristics. No animal is completely human, but each one exemplifies a particular trait. The fox is always cunning, the lion is strong, the wolf is rapacious, the dog is obedient, ants are industrious, and so on. Aesop may have been the first European storyteller to derive moral lessons from the supposed behavior of animals, but his fables succeeded only because people had always attributed human traits to particular animals. It is almost impossible not to regard household pets as almost human and to see human characteristics in other species. Many everyday expressions reflect this. Someone in a bad temper is "like a bear with a sore head." An especially nimble thief is a "cat burglar." A person who refuses to see danger that is obvious to everyone else is said to be burying his head in the sand—"like an ostrich." A deceitful person is "a snake in the grass" and one who is working hard and persistently at a difficult task is said to be "beavering away."

Metaphors apart, animals that talk and give moral advice are nowadays confined mainly to fairy tales and other children's stories, but in the Middle Ages grown-ups took them seriously. In those days, animal stories illustrated Christian teaching, and this chapter tells a few of those stories. It begins by explaining the context in which the stories arose and ends more recently, with the difficulties zoologists faced when for the first time they encountered and attempted to classify an animal that refused to fit into any of their existing categories.

MORAL LESSONS TO BE DRAWN FROM ANIMAL BEHAVIOR

A modern book or article describing an animal would include pictures, probably photographs, and would reveal the animal's anatomy, ecology, and way of life. Obviously, the text and illustrations would be simpler in a book intended for young readers than in an undergraduate textbook, and an original research paper would be more complicated still. But all of these would have the same objective—to inform the reader about the animal. They would differ only in the vocabulary they employed and the amount of knowledge the authors assumed their readers already possessed.

People in the Middle Ages were no less curious about animals than people living today, and those who could afford them loved illustrated books. Authors, however, approached their task from a very different point of view. They were not concerned to explain anatomy or behavior. They had a much more serious purpose. They knew, as all Europeans knew, that God had created all of the animals they were describing, and God did nothing without a purpose. Every animal had been placed on Earth for a reason. God created some animals to serve humans, but in addition all animals, including wild species, existed in order to instruct mortals as a means of guiding them toward salvation. All humans were believed to be sinful, and, by studying the nature with which God had imbued each animal, the sinner might learn how to repent and reform. The resulting book of animals was, therefore, much more a work of theology and moral instruction than one of zoology.

Most of the lessons are simple. Good Christians should emulate the industry of the ant, for instance, and the purposefulness of the bee, and they should avoid the hypocrisy of the crocodile that lives as a land animal during the day and inhabits the water at night. Behind the simple morality, however, there was a more subtle symbolism revealed in the scriptures, where a particular animal might be either good or evil depending on the biblical context. The author had to understand and explain the mystical symbolism as well as providing moral instruction.

Writing a popular book about animals in 12th- and 13th-century Europe was quite different from writing a book about the same animals in the 21st century, but it was no easier. The medieval author had to consult as many scholarly sources as a modern author, but they were not the same kinds of sources because the author had a different purpose.

BESTIARIES

In the Middle Ages people loved animal stories throughout Europe, the Middle East, and North Africa, and the stories they enjoyed were always intended for Jewish, Islamic, or Christian religious instruction. Jews, Christians, and Muslims all derive part of their teachings from the Old Testament, and it is there that believers are advised to learn from nature, for example in the following verses from the book of Job:

> But ask now the beasts, and they shall teach thee;
> And the fowls of the air, and they shall tell thee:
> Or speak to the earth, and it shall teach thee:
> And the fishes of the sea shall declare unto thee.
> Who knoweth not in all these
> That the hand of the Lord hath wrought this?
> In whose hand is the soul of every living thing,
> And the breath of all mankind. (Job 12:7–10)

A medieval illustrated book of animal descriptions was called a *bestiary.* They reached the peak of their popularity during the 12th and 13th centuries. Most monastic libraries held at least one bestiary. Few ordinary folk were literate, but with the monks and priests to help interpret them they could understand the pictures. One of the most complete bestiaries to survive is held at the University of Aberdeen, Scotland, and the Aberdeen Bestiary states its purpose clearly. This is "to improve the minds of ordinary people, in such a way that the soul will perceive physically things which it has difficulty grasping mentally: that what they have difficulty comprehending with their ears, they will perceive with their eyes."

The authors of the bestiaries did not confine themselves to the animals they could see around them in the streets and fields. They also described animals they had probably never seen, such as elephants, crocodiles, and hyenas, as well as animals like the leucrota that were entirely mythical. The leucrota was said to live in India. Fierce and swifter than all other beasts, it was the size of an ass and had the hindquarters of a stag, the chest and legs of a lion, the head of a horse, its hoofs were cloven, and its mouth stretched from ear to ear, with a single continuous bone in place of teeth. Its voice imitated human speech. No lesson is drawn from the leucrota or from most of the other exotic animals. They are probably included for entertainment, and many of them also appeared in illuminated manuscripts, maps, and heraldic devices.

Medieval writers based their work on that of earlier authors and did not aim for originality, which they would have considered a departure from the established truth. They took many of their accounts from Greek and Roman sources. The description of the leucrota, for instance, came from Pliny the Elder's *Natural History* (see "Pliny, Collector of Information" on pages 21–26). Their most important source, however, was a text that is known simply as the *Physiologus.*

The *Physiologus* was written in Greek by someone living in Alexandria, probably in the second century C.E., but possibly in the third or fourth century. No one knows who wrote it. *Physiologus* was not its title. It is called that because each of its stories begins with the words "the physiologus says" and *physiologus* is usually translated as "the naturalist," although this is not a work of natural history. Its stories are clearly Christian. It tells of a lion whose cubs are born dead, but come to life when the old lion breathes on them, and of the phoenix that burns itself to death and rises from the ashes three days later. Both of these refer to Christ. The unicorn that can be captured only in the lap of a virgin refers to the incarnation, and the pelican that sprinkles its own blood over its dead offspring in order to bring them to life refers to the crucifixion. The hoopoe is a bird that cares for its elderly parents, just as Christian children should care for their parents.

The *Physiologus* was translated into Latin in about 400, into Aethiopic at some time in the fifth century, and later into German, Armenian, French, Syriac, and Arabic. By the 13th century the Latin text had been translated into most European languages, including English and Icelandic. The Greek version spread through Eastern Europe and was translated into Slavic and from Slavic into Romanian and other languages. The bestiary authors worked from the Latin translation, however, and bestiaries were beginning to appear by the 10th century. These contained stories from the *Physiologus* mingled with other stories from *Etymologiae,* a work by St. Isidore (ca. 560–636), an archbishop of Seville.

HOW DOTH THE LITTLE BUSY BEE

The tradition established by the bestiaries lasted for many centuries, and moralizing teachers continued to write stories and verses exhorting children to good behavior and industriousness. None of these exhortations is more famous, perhaps than "Against Idleness and Mischief," which appeared in *Divine Songs for Children,* written by the English hymn writer Isaac Watts (1674–1748). That is the poem warning that:

> Satan finds some mischief still
> For idle hands to do.

In true bestiary style, the poem uses the bee as an example, commencing with the following stanza.

> How doth the little busy bee
> Improve each shining hour,
> And gather honey all the day
> From every opening flower!

The following stanzas describe the way the bee works, and the poem ends with the exhortation:

> In books, or work, or healthful play,
> Let my first years be passed,
> That I may give for every day
> Some good account at last.

By the 19th century the little busy bee was looking absurd, and Lewis Carroll (Charles Lutwidge Dodgson [1832–98]) parodied it in *Alice's Adventures in Wonderland,* as:

> How doth the little crocodile
> Improve his shining tail . . .

The bestiaries, more robust and less sanctimonious than Isaac Watts, held up the bees as an example of an ideal society. Only the bees, among all creatures, share everything, working and living together in harmonious collaboration and exhibiting great skill in the way they make honey and construct their houses. They choose a king—the queen bee was then unknown—and although they are subject to their king they remain free, because they recognize him as their elected leader and are faithful to him. No bee dares leave the hive in search of food unless the king has gone ahead and leads the flight. The king possesses a sting, but refrains from using it to punish wrongdoers. Instead, offenders must turn their stings on themselves. Bees, say the bestiaries, are known by observation to be born from the corpses of cattle as worms that later turn into bees. The insects born in a similar way from horses are hornets, those from mules are drones, and those from donkeys are wasps.

The bestiary account of bees was taken from Pliny and other Roman writers. Pliny was especially enthusiastic. In Book XI of his *Natural History* he wrote the following about bees:

> They are patient of fatigue, toil at their labours, form themselves into political communities, hold councils together in private, elect chiefs in common, and, a thing that is the most remarkable of all, have their own code of morals.

Despite Pliny's assertion, insects are not intelligent in the way humans are intelligent, but they do communicate. The Austrian zoologist Karl von Frisch (1886–1982) won a share of the 1973 Nobel Prize in physiology or medicine for his discovery of the *dance language* that bees use (see "Karl von Frisch and the Dances of the Honeybee" on pages 176–179). Bees may not be as politically and morally sophisticated as Pliny and the bestiary authors believed, but their skills at communication and navigation are nevertheless remarkable.

HEDGEHOG

European hedgehogs (*Erinaceus europaeus*) are popular animals and useful to gardeners because they eat slugs and snails. They are insectivores, feeding on insects and other invertebrate animals, and they are quite common although, being nocturnal and shy, they are not often seen. Over the centuries, their reputation has changed dramatically, and the medieval bestiary writers saw the inoffensive hedgehog in a very different light.

The hedgehog's most obvious feature is its protective coat of spines. When threatened, the hedgehog rolls itself into a tight ball with the spines on the outside. This is a highly effective defense against most aggressors (although a fox may be able to unroll it). But

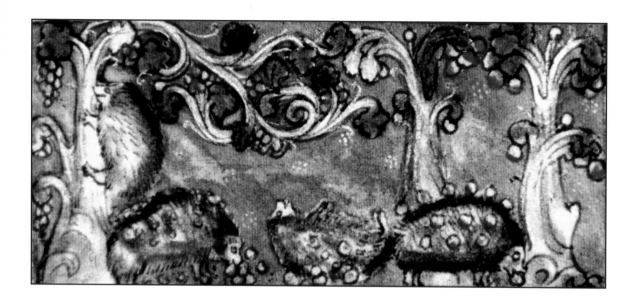

This illustration from a 13th-century bestiary shows hedgehogs climbing a vine in search of grapes. *(The Granger Collection)*

the hedgehog, which one bestiary describes as the most timid of animals, reinforces that protection by hiding among rocks. In this it is like the man, bristling with sins, who fears the coming judgment and finds a secure refuge in the rock of Christ.

So the hedgehog was seen as a prudent animal, but its sins were serious. Medieval authors believed that hedgehogs climb vines—or apple trees in some versions—and pick the fruit. The hedgehog then rolled onto the grape or apple, impaling it on its spines, and in this way it was able to carry the fruit home to its young, waiting in its den. The illustration above, from a bestiary compiled in the 13th century, shows hedgehogs climbing a vine to take the grapes. This was behavior typical of Satan. The hedgehog cheated others of their fruits, as the Devil robs virtuous people of their spiritual fruits.

Its technique for feeding its young was only one aspect of the hedgehog's ingenuity. It was also believed to be very skillful in the way it constructed its den. This was said to have two holes for ventilation. When the hedgehog noticed that the wind had started to blow from the north—a northerly wind is a cold wind in the Northern Hemisphere—it quickly blocked that hole. There are dangerous lands to the south of Europe, and, when the wind blew from that direction, driving mists before it, the hedgehog blocked that hole and ran for its supply of fresh air to the hole facing north. This behavior suggested to the bestiary writers that the hedgehog is a sinner, full of vices like spines, and skilled in wicked cunning, deceit, and robbery. The hedgehog was a sinful animal from whose example Christians might learn to be honest and open.

FOX

The red fox (*Vulpes vulpes*) is renowned for its cunning. The bestiary writers believed that its Latin name, *vulpis*, came from *volupes*, which means "twisty foot" and referred to the supposed fact that a fox never runs in a straight line. (In fact, foxes do run in straight lines when being pursued, although not when foraging for food.) An animal that follows a crooked path is clearly devious, and that was part of the fox's cunning.

It was only the start of the animal's deviousness, however. When a fox was hungry and could find nothing to eat, the bestiary writers believed that it would roll itself in red-colored mud, so its coat became spotted with red, resembling blood. It would then lie quite still, its eyes closed, its tongue lolling, and holding its breath, as though it were dead. The birds, seeing what they thought was a dead fox, would pluck up their courage and approach it to feed, but the instant they began to peck the fox would leap up, seize the birds, and devour them. In that, the fox resembled Satan, who also seems to be dead until an unwary sinner approaches too close and is seized by the throat. But for holy men and women Satan is truly dead, for their faith reduces him to nothing.

WEASEL

The European weasel (*Mustela nivalis*) is tiny. Males, which are bigger than females, are only 8–9 inches (21–23 cm) long, not counting the short tail. It is also a fearless predator, hunting rabbits and other animals much larger than itself, and it will launch a ferocious attack on any human who corners it. In the Middle Ages, it was thought to be a long-bodied mouse, and it was known to attack snakes and mice and was the only animal capable of defeating the fearsome basilisk (see "Basilisk" on page 85).

Weasels were partly domesticated and kept indoors to control rodents. Whether domesticated or wild, it was well known that shortly after giving birth, a weasel would carry its young to another place, moving them several times and each time to somewhere different. This was seen as very cunning behavior. Its method of reproduction was also unusual, to say the least. Some authorities the bestiary writers used as sources held that the weasel conceived through its ear and gave birth through its mouth. Others believed the opposite: that they conceived through the mouth and gave birth through the ear. Weasels also possessed remarkable healing powers. If a weasel discovered that its young had been killed, it was able to bring them back to life. There were thought to be two types of weasels. One lived in the open and close to people, and another, larger type, lived in the forests. The larger animal was, in fact, the closely related stoat, also called ermine (*M. erminea*).

The behavior of the weasel reflected that of the person who listened readily to the word of God, but who was held back by love of worldly things, and so rejected all that they had been taught and lost all interest in learning more. The weasel also symbolized a thief.

PELICAN

Pelicans are a family (Pelecanidae) comprising eight species of fairly large, gregarious birds that feed on fish, which they scoop from the water while flying low over the surface. To the medieval mind, however, the pelican was a very different bird, rich in symbolism, and the pelican of the bestiaries bears no relation to the real bird. As well as appearing in most bestiaries, the pelican was widely used as a heraldic device, appearing on banners, shields, and coats of arms.

According to the medieval authors, the pelican was an Egyptian bird that inhabited remote areas near the River Nile. It showed very great devotion to its young, but, alas, its young were ungrateful. They became disobedient and rebellious, sometimes even striking their parents in the face, until at last the parents could endure it no more and lost their tempers. In their rage they killed their young, and immediately began to mourn their death. After three days of mourning the mother pelican (or in some versions the father) pecked open her own breast, pouring her blood over the bodies of her young. The parent's blood, and the love with which it was shed, brought the young back to life. The image of the pelican with blood

Accendat·in·nobis
dominus·ignem
sui·amoris·et
flammam·aeternae
charitatis

This engraving by the Welsh artist David Michael Jones (1895–1974) was made in 1929. It shows a pelican in her piety, feeding her chicks with her own blood. *(Wolseley Fine Arts, London)*

pouring from her breast over her chicks is known as "the pelican in its piety," and it was one of the most popular images in medieval Christendom. The illustration above shows a modern depiction of the pelican in its piety. Like the medieval pictures, this engraving shows that the chicks appear to be very lively. That is because the blood revives them instantly. Some medieval depictions show other stages in the story, either of the young attacking their parent or of the enraged parent killing the young.

The symbolism is obvious. God created men and women, but they reject and abuse him. Christ then suffered crucifixion, shedding his blood to bring salvation and give humans eternal life. The pelican in its piety therefore symbolizes the passion of Christ.

EAGLE

Eagles—there are many species—are charismatic birds, symbolizing power, vigilance, freedom, and the capacity to launch a swift and unhesitating attack. When air forces design badges for their aircrews they use wings and the wings are, or are meant to be, those of eagles. The imperial eagle was the symbol of the Roman Empire, and eagles, sometimes double-headed, are still used as national emblems in Albania, Armenia, Austria, Egypt, Germany, Mexico, Panama, Philippines, Poland, Romania, Russia, Serbia, and, of course, the United States. The illustration at left shows the coat of arms of the Russian Federation, dominated by the double-headed eagle.

Medieval writers were no less impressed by the eagle, and especially its keen eyesight and sharp bill. Indeed, the bestiary in the Bodleian Library at Oxford University states that: "[T]he eagle is so called because it is eagle-eyed." The bestiaries described it flying so high that it was out of sight to people on the ground, but

The coat of arms of the Russian Federation is dominated by a double-headed eagle.

even so able to see a fish swimming near the surface of the sea. It would then swoop down—like a thunderbolt according to one bestiary—and seize the fish.

Its eyesight was so important to the eagle that it subjected its young to a stern test. One by one it took them in its claws and lifted them from the nest and into the full glare of the Sun. Chicks that averted their gaze were rejected and dropped as though they were of an inferior species that had secreted its eggs in the eagle's nest and were unworthy of so great a father. The eagle accepted as its own true offspring only those young birds that gazed unblinkingly at the Sun.

As the eagle grew older, however, its eyesight began to deteriorate, its wings weakened and grew heavy, and its bill became long and cumbersome. The eagle then found a fountain and from there it flew high into the sky, approaching so close to the Sun that the heat burned the mistiness from its eyes and its feathers scorched. It then returned to the fountain, diving into the water three times. This restored its wings to their full strength, and with its eyes bright once more the eagle was fully recovered except for its bill, which it chipped and ground into a proper shape on stones.

The message for men and women was that as they grew old, with old and tattered clothes and the eyes in their hearts and minds darkened, they should seek the spiritual fountain of the Lord and the church and be reborn by baptism in water. In the words of Psalm 103, verse 5, "So that thy youth is renewed like the eagle's."

ELEPHANT

There is a famous public house on the southern side of the Thames in London that gives its name, Elephant and Castle, to the surrounding district. Opinions differ concerning the origin of the name, some holding that it is a corruption of "Infanta of Castille." However, in medieval times, the depiction of an elephant with a war-tower on its back, known as an elephant and castle, was very common in heraldry and church carvings. The elephant was of great interest to the writers of bestiaries, who noted that: "There is no beast greater than this. The Persians and Indians put wooden towers on his back and fight with arrows as if they were on top of a wall."

The bestiaries describe elephants as gentle animals with a keen intelligence and excellent memory. They move around in herds, and although an elephant strikes terror into bulls it is terrified of mice. Elephants have no joints in their knees. Consequently, if an elephant falls down it is unable to rise again. It sleeps leaning against a tree and Indian hunters catch elephants by cutting partway through the tree so that when the elephant leans against it the tree falls and the elephant with it. As it falls, say the bestiaries, the elephant trumpets loudly and at once a huge elephant appears on the scene and tries to lift him, but is unable to do so. Both elephants then trumpet, summoning 12 elephants, but their combined efforts are still insufficient, so they all trumpet together. This time a little elephant appears, puts its trunk beneath the fallen elephant, and lifts him. The little elephant is very special. Nothing evil, not even a dragon, can do harm in any place where some of its hair and bones have been burned.

The large elephant symbolized Hebrew law. It could no more lift the fallen elephant than the priest and Levite could help the man who fell among thieves and left him half dead in Luke's gospel (Luke 10:30–32). The 12 elephants symbolized the prophets of the Old Testament. None were able to raise the fallen elephant but, like the good Samaritan (Luke 10:33–35), the little elephant accomplished the task just as Christ came to Earth to save man, who had fallen, and raise him up.

Elephants were said to live for 300 years, and in the course of that long life a female gave birth only once and to a single offspring. They were reluctant to mate, but when they wished to produce young, the couple traveled eastward to a place close to Paradise, where a tree called the mandragora grew. The female picked a fruit from the tree, gave it to the male, and persuaded him to eat it. After they had both eaten of the fruit they mated, back to back, and the female conceived at once. Her pregnancy lasted for two years. When the time came for her to give birth the female entered a pond, while the male stood guard to protect her from the elephant's only enemy—dragons.

The male and female elephants represented Adam and Eve. Until they sinned they knew nothing of mating, but when they ate of the fruit of the mandragora tree the male acquired understanding and the female at once became pregnant. As soon as this happened the two elephants had to leave Paradise.

Male elephants never fought over females, because adultery was unknown to them and, in the words of the bestiaries, "the goodness of mercy is within them." When they came across men who were lost in the desert elephants led them back into familiar ways. If they encountered a flock of sheep they would protect it so that no missile could kill any of them, and if they were made to fight in a battle they always took great care of the wounded and exhausted.

VIPER

Married couples should remain faithful to each other and show tenderness and mutual understanding. Failure to do so may expose them to a fate similar to that of the viper. The vipers comprise a family (Viperidae) of about 100 species of venomous snakes found in Europe, Asia, and Africa. The king cobra (*Ophiophagus hannah*) of Asia is the largest member of the family and the largest of all poisonous snakes, sometimes growing to more than 16 feet (5 m) in length. Many viperid snakes are called vipers, but the common viper or adder (*Vipera berus*), which occurs throughout Europe as far north as the Arctic Circle, is the only venomous snake native to Great Britain (and it is absent from Ireland). It is not aggressive and its bite, though it can be painful, is seldom fatal to human adults.

That is not how the viper is portrayed in the bestiaries. There it is described as a lustful animal, the most evil of all creatures, and the most cunning of all serpents. In order to indulge its sexual needs the viper goes to the seashore and hisses to invite an eel to approach and mate with it. The eel does so willingly.

The viper needs this unnatural liaison because its method of reproduction is invariably fatal. When vipers mate, according to the bestiaries, the male places his head inside the female's mouth. The female then bites off the head, killing the male. The young vipers then grow inside the female's body, but when the time approaches for them to be born, instead of waiting for a normal birth they bite their way through the mother's body, killing her. Thus reproduction kills both the male and the female. "O man, learn from this," says the Bodleian bestiary, "anyone who tries to seduce another man's wife may well embrace a snake." It is true that the common viper gives birth to live young, by *ovovivipary,* but no other part of the bestiary account is true.

CROCODILE

Crocodiles were understood by the writers of bestiaries to live in the River Nile and to be four-footed animals that were yellowish in color.

The crocodile, as it is portrayed in the Worksop Bestiary, written in England, perhaps in Lincoln, in the 12th century. The crocodile is shown living in the River Nile and devouring a man. *(The Pierpont Morgan Library/Art Resource, NY)*

The Bodleian bestiary suggests that the crocodile's name referred to the color of crocuses. It was equally at home on land or in the water, entering the water at night and resting on land by day, and it was more than 20 cubits (about 30 feet; 9 m) long. The crocodile was the only animal that could move its upper jaw while holding the lower jaw still. It was armed with huge teeth and its skin was so tough that stones thrown at it caused it no injury. Only one animal could kill it, and that was a particular fish with serrated spines that could rip open the crocodile's belly.

Famously, the crocodile was believed to weep (crocodile tears) in feigned contrition after devouring a human. The illustration above, from a medieval bestiary, shows a crocodile—with long, straight legs and a long tail—in the act of killing a human. The artists who illustrated bestiaries had never seen a crocodile or many of the other animals they were expected to draw and had to rely on written descriptions and their own imaginations. It is hardly surprising, therefore, that bestiary illustrations often bore little resemblance to real animals.

Crocodile dung was used to make an ointment. When old women smeared the ointment on their faces their wrinkles vanished and they looked young again, retaining their youthful appearance until their sweat washed the ointment from their faces. If a man ate the ointment it would make him weep. The effect of the ointment was reminiscent of the way inexperienced persons often praised the wicked, their praise being like a salve that made bad deeds appear as acts of heroism. But when a judge struck the wicked for their misdeeds, all the splendor of that praise vanished.

The crocodile represented the hypocrite. Its habit of living in the water by night and on land by day symbolized the person who lived a depraved life out of sight in water and darkness, while enjoying a holy and upright reputation during the day on land, where others could see it. Its ability to move its upper jaw independently of its lower jaw was also a sign of hypocrisy, suggesting those who presented the church's teachings to others, but failed to practise what they preached.

BARNACLES

Frederick II of Hohenstaufen (1194–1250), the Holy Roman Emperor and king of Sicily and Jerusalem, had a rational outlook on the world and was an authority on falconry and bird behavior. When presented with beliefs that struck him as dubious he tried to examine them scientifically. That was the spirit in which he sent an expedition into the far north of Europe to investigate the reproduction of the barnacle goose. The archdeacon of Brecon, Gerald of Wales (ca. 1146–ca. 1223), also known in Welsh as Gerallt Cymro and in Latin as Giraldus Cambrensis, had recently summarized in his book *Topographia Hiberniae* (Description of Ireland) the opinion widespread among medieval scholars that barnacles were rather peculiar birds. The bestiaries repeated Gerald's account, as in the following excerpt:

> In Ireland there are many birds called barnacles, which nature produces in a way which contradicts her own laws. They are like marsh geese, but smaller. They first appear as growths on pine-logs floating on the water. Then they hang from seaweed on the log, their bodies protected by a shell so that they may grow more freely; they hold on by their beaks. In due course they grow a covering of strong feathers and either fall into the water or change to free flight in the air. They take their food from the sap of the wood and from the sea in a mysterious way as they grow. I have seen them myself, with my own eyes, many times: thousands of these small birdlike bodies hanging from just one log on the seashore, in their shells and already formed.

Gerald drew a clear lesson from this: "In some parts of Ireland, bishops and men of religion eat them during times of fasting without committing a sin, because they are neither flesh, nor born of flesh."

The goose barnacle (*Lepas anatifera*) is a crustacean that attaches itself to rocks or pieces of driftwood and feeds by filtering particles from the water. Goose barnacles were once thought to develop into barnacle geese (*Branta leucopsis*).

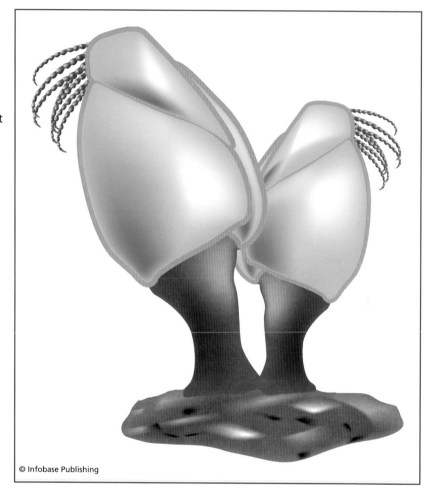

© Infobase Publishing

Thus it became acceptable to eat barnacle geese on days when the eating of meat was forbidden. The bestiaries repeated this lesson and added a second, which was that nature produced the barnacle goose without a male or female in order to instruct people and confirm them in their faith. The first human was made from clay, the second was born of a male but no female, and only the third was born of male and female. The fourth generation—Christ—was born of a woman alone.

When Frederick's envoys returned from the north they brought with them the shells attached to rotten wood, but these bore no resemblance to any bird. The German scholar St. Albertus Magnus

(Albert the Great, ca. 1206–80) also investigated this story. He and some of his friends bred a barnacle goose with a domestic goose and found that the female laid eggs that hatched into perfectly ordinary goslings. Despite this, presumably relying on the authority of the bestiaries and the testimony of Gerald, the belief persisted as late as the 16th century that barnacle geese spent part of their life cycle as shellfish.

The confusion—for such it was—arose from the fact that the barnacle goose (*Branta leucopsis*), which is a small black-and-white goose and most definitely a bird, visits the coastal regions of Ireland and western Scotland in winter. In spring it migrates northward to breed in the Arctic. Consequently, although the goose was familiar in winter, it was never seen to breed, and in medieval times scholars did not know that certain birds migrate, breeding in one place and spending the remainder of the year in another. It sometimes happened in summer, however, that a storm washes ashore pieces of driftwood to which goose barnacles (*Lepas anatifera*), also called gooseneck barnacles, are attached. As the illustration at top of page 76 shows, the white plates of a goose barnacle are vaguely reminiscent of the plumage of a goose and the way its "head" sways in the flowing water is somewhat birdlike. At all events, that is how the two entirely unrelated species came to be thought of as one.

Goose barnacles are crustaceans that attach themselves by long stalks to solid surfaces such as rocks or pieces of wood. The main part of the barnacle's body is enclosed between rigid plates, giving it a roughly triangular shape resembling a head. The barnacle feeds by opening its plates and extending tentacles that filter particles from the water.

CAPTAIN HUNTER, THOMAS BEWICK, AND THE DUCK-BILLED PLATYPUS

On April 19, 1770, an expedition led by James Cook (1728–79) on board HM Bark *Endeavour* came in sight of the Australian coast. When the expedition scientists went ashore and began to examine the flora and fauna, they discovered species that were entirely unknown in Europe. They met the kangaroo, giving it the name local people used when the scientists asked them what they called the animal. In the years that followed, as more and more specimens of the Australian fauna arrived back in London, it became evident that although the species were new, they were nevertheless related to species that were already known. This strongly suggested that all animals are descended from a much smaller pool of common ancestors. But their confidence was about to be shaken.

At first, the species being discovered were new and strange, but they were not bizarre. All of them fitted into the broad categories by which animals were being classified. Then, on August 15, 1798, Captain John Hunter (1737–1821), a former British naval officer and since 1795 the governor of New South Wales, wrote to the naturalist Sir Joseph Banks (1743–1820) in England to tell him he was sending him some specimens that he asked Banks to pass on to the Literary and Philosophical Society of Newcastle upon Tyne, of which Hunter was an honorary member. The most important of the specimens was an animal Hunter called the womback (wombat). He also added that: "In the same cask [as the womback] will be found the Skin of a Small Amphibious Animal of the Mole kind which boroughs (*sic*) in the Banks of fresh water lakes." He could send only the skin because the weather had been very warm, making it impossible to preserve the whole animal in spirits, but he enclosed a drawing in his letter. The specimen reached the society late in 1799 and was presented together with Hunter's drawing at a meeting of the society held on December 10.

The animal "of the mole kind" was the first platypus to be seen in Europe, and the naturalists did not know what to make of it. At first they suspected it was a fake, made by stitching together a duck's bill and body parts from several mammals. Closer investigation showed it to be genuine. It was a small animal—males grow up to about 24 inches (60 cm) long and females are smaller—but what kind of animal?

Thomas Bewick (1753–1828) the famous engraver and naturalist, who was a native of Newcastle, reproduced Hunter's drawing in the fourth edition of his *History of Quadrupeds* and added a description: "It appears to possess a three fold nature, that of a fish, a bird, and a quadruped [the usual term for a mammal], and is related to nothing hitherto seen." It had a bill resembling the bill of a duck, suggesting it fed in the same way, very small eyes, four short legs, webbed feet, and long, sharp claws. George Shaw (1751–1813), the assistant keeper in the natural history department

DUCK BILLED PLATYPUS

PL.36

of the British Museum, published the first scientific description of the animal in his *The Naturalist's Miscellany*, published in 1799. He also named it *Platypus anatinus*—duck-billed platypus—but was unable to classify it. The platypus has retained its common name, but in 1800 the German anatomist Johann Friedrich Blumenbach (1752–1840) altered its generic name to *Ornithorhynchus*; it is now *Ornithorhynchus anatinus.* The illustration above is from an engraving published in 1811.

A colored engraving of a platypus (*Ornithorhynchus anatinus*) that appeared in *The Natural History of Quadrupeds and Cetaceous Animals,* published in 1811 *(Time Life Pictures/Stringer)*

   Further investigation revealed that the platypus lays eggs, secretes milk to feed its young but has no mammary glands, and possesses a spur on each of its ankles that it can erect to inject venom. Was it a bird, a reptile, or a mammal? Just when European naturalists had devised a classification into which every vertebrate animal could fit, this strange little beast threatened to demolish the entire edifice. Sir Everard Home (1756–1832) published the first study of its anatomy in 1800, demonstrating that the similarity between its bill and that of a duck was purely superficial. Behind its bill, the platypus has cheek pouches with horny ridges in which it stores food while it is chewing. Platypuses possess teeth when they hatch, but lose them by the time they reach maturity. George Bennett (1804–93) published the first description of one living in its natural habitat, and the next account of a live platypus did not appear until 1927, written by the Australian naturalist Henry Burrell (1873–1945).

The platypus does not feature in any medieval bestiary, but to the naturalists tasked with defining it, this small animal—certainly not of the mole kind!—must have seemed as fantastic as any beast described at 10th hand from travelers' tales. The matter has been resolved, of course. *Ornithorhynchus anatinus* is the only member of the family Ornithorhynchidae in the order Monotremata. It is a mammal that retains many primitive features, the most important being the common opening, called a *cloaca,* to the rectum and urinogenital system.

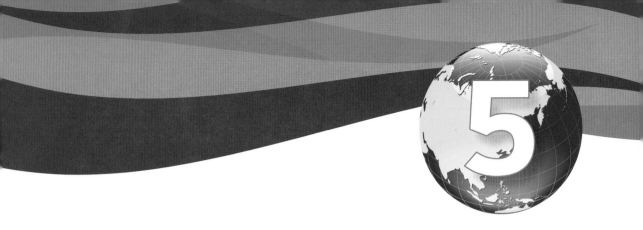

Fantastic Beasts

Most medieval storytellers and illustrators were not zoologists, nor did they pretend to be. They aimed either to entertain their readers or to provide them with religious and moral instruction, and sensationalism served both ends very well. The entertainers knew, just as well as any present-day newspaper editor or owner of a TV station, that people love to be amazed and frightened. But the medieval entertainers had a distinct advantage, because their readers had no way of checking the veracity of what they were told. Writers could not exaggerate the dangers posed by familiar animals—not even really fierce ones such as bears and wolves—simply because they were familiar and people knew how to live with them. But a fire-breathing dragon was a very different proposition. No reader had ever met with one, but the very idea that such an encounter might be possible was enough to be frightening—and there was no way to prove that it could not happen.

Moralists had a different purpose. They sought to draw parallels between animal characteristics and behaviors and those of humans. They did not describe nonhumans as though they were humans—a practice called *anthropomorphism* that can seriously mislead students—but compared animal traits with those of wicked or virtuous people. In order to make this work, they were compelled to describe animal traits that in fact do not exist, so either the information on which they based their reports was mistaken or they made it up.

Medieval authors and artists did not travel far from home. They learned about the animals they described and painted from merchants, sailors, soldiers who had fought in overseas campaigns, and other seasoned travelers. This was the obvious thing to do, but the approach was seriously flawed. A modern scientist studying wildlife must be single-minded in pursuit of that goal and be prepared to visit remote places and spend many hours waiting patiently for animals to appear. A tourist or person on a business trip has very little chance of seeing local wild animals, far less of examining them closely, and a sailor who visits only busy seaports or a soldier moving with a large army has even less chance. So most of the people supplying the accounts that found their way into books about animals had never actually seen those animals. They were repeating what they had heard from others, who were also repeating stories that had passed from mouth to mouth. By the time they reached Europe, these travelers' tales had the status of modern urban myths.

Amid the distorted accounts of real animals, returning voyagers also brought descriptions of animals that were entirely fictitious. This chapter describes a few of these—including the dragon—and the lessons that were drawn from them.

MANTICORA

The manticora or manticore was said to live in India. It had the head of a man but with three rows of teeth set alternately, a body of a lion, and a tail with a sting like the sting of a scorpion. Some writers described its voice as hissing; others said it was high-pitched piping.

Clearly it was an alarming beast, but its habits were worse. Despite its human head, the manticora had an insatiable appetite for human flesh. The manticora was also extremely agile. It was so sure-footed and its feet and legs were so strong that it was able to jump over any ditch or other obstacle designed to hold it captive.

The manticora first appeared in Persian literature, and its name is derived from two Middle Persian words, *martya* and *xwar*, meaning "man" and "to eat." The Greeks pronounced the name as *mantikhoras*, and both Aristotle and Pliny wrote about it. More recently, the name manticora has been given to a genus of nocturnal, flightless, and fiercely predatory tiger beetles that have very large *mandibles*—the pair of mouth-parts used to seize prey.

THE MANTICORA

The bestiaries describe the manticora as having gray eyes and a body that is blood-red from the neck down. Many of the illustrators ignored the coloring, however, and few of them attempted to show either the three rows of teeth or the scorpion like sting at the tip of the tail. The illustration shown above, from a book published in the 19th century, is typical. It gives the beast a head of neatly cropped hair, a beard, and a flowing lion's mane, with a blue body and a tail that ends in a mass of spikes forming a club.

The manticora, with its human head, lion's body, and a scorpion sting in its tail, featured in many medieval travel books and bestiaries. This example is more recent. It appears in the 1886 edition of *Myth-Land* by F. E. Hulme. *(Charles Walker/Topham/The Image Works)*

DRAGONS

The dragon was the only animal that could overwhelm and kill an elephant. The largest of all serpents, indeed the largest of all animals, it would wrap its body around its victim and then strike repeatedly with its tail until the elephant was exhausted. Finally, it killed by constriction.

The dragon that is the national emblem of Wales.

Tales about dragons occur throughout the world in almost every culture, and some dragons have been adopted as national emblems. The illustration above shows the Welsh dragon, which appears on the Welsh flag. This red dragon fought and defeated a white dragon in a battle watched by the Welsh king. Merlin the magician explained to the king that the red dragon symbolized the Welsh and the white dragon the Saxons, implying that one day the Welsh would defeat the English. Dragons appeared in heraldry and on the shields of medieval knights as symbols of martial strength, a reputation based on their size and power.

In the bestiaries, in contrast, the dragon is pure evil. They describe it as having wings, a crest, a small mouth, and narrow nostrils. It extends its tongue, but its strength lies not in its jaws and teeth, but in its tail. It possesses no venom, but has no need of venom because it kills anything it embraces. The dragon lives in caves in India and Ethiopia, where it finds the heat it needs. When it emerges from its cave the motion shakes the air, making it glow and shimmer with the heat.

The dragon is the fairest of serpents and symbolizes the devil, who also rises from below and makes the air glow as he transforms himself into an angel of light to deceive fools with promises of pleasure and vainglory. The dragon has a crest because the devil is proud, and its strength is in its tail rather than its teeth because the devil, having lost his power, can achieve his ends only through deceit and lies. It waylays elephants just as the devil waylays those proceeding toward heaven. The devil suffocates his victims with sin and despatches them to hell.

BASILISK

The dragon may have been the largest of all serpents, but the king of the serpents was the basilisk, sometimes called regulus, which means little king. It was small, no more than six inches (15 cm) long, with white spots on its body, but it was fearsome. Those who saw the basilisk fled from it, because its scent would kill them and, should anyone attract its attention, it could kill by merely looking at someone. Its bite was also lethal and, according to some accounts, even the sound of its hissing could kill. If a bird should fly past the basilisk, its fiery breath would kill it in flight, and the basilisk would then swallow it. The basilisk lived in dry places, like the scorpion, but sometimes it came close to water. When it did so, it would poison the water so that anyone drinking it contracted hydrophobia and died in panic. The illustration of the basilisk on page 86 is from a drawing made in 1512 by the German artist Albrecht Dürer (1471–1528).

Some people thought they might kill a basilisk by looking at it before it had a chance to look at them, but no one really trusted this idea. Alternatively, it might be killed by holding up a glass container that would reflect the basilisk's glance back on itself. This suggestion was also dubious.

There was one certain way to destroy the basilisk, however, and that was to employ a weasel. Weasels were known to be immune to basilisk venom. Place a weasel down the hole where a basilisk was hiding and the basilisk would flee, but the weasel would pursue and eventually catch and kill it. This was because, in the words of the Bodleian bestiary, "the Creator of all things has made nothing for which there is not an antidote."

The basilisk was a serpent that might have been hatched by a toad from an egg laid by an elderly cockerel. This drawing made in about 1512 by the German artist Albrecht Dürer (1471–1528) is called "The Sun, the Moon and a Basilisk." *(The British Museum/Topham-HIP/The Image Works)*

The origin of the basilisk is uncertain, but the idea emerged in the Middle Ages that it might have been hatched by a toad from an egg laid by an elderly cockerel. This led to its alternative names of basili-coc and cockatrice.

The basilisk signified the devil, who kills the careless sinner with his venom. The devil is conquered by: "the soldier of Christ who puts all his hope in the Lord, whose power overcomes and tramples underfoot all hostile forces."

ONOCENTAUR

The centaur of classical mythology had the upper body of a man and the lower body of a horse. The onocentaur, or onoscentaurus, is a close relative, being half man and half donkey. Like the centaur, he is often depicted in bestiaries holding a bow or club, and the bestiaries always link him with the siren. In one bestiary the onocentaur and siren are shown in the sea together, apparently attacking a ship.

Stories about the onocentaur first appeared in ancient Greece, and it was believed to live in Africa. Its human arms served a dual purpose, because when it needed to move fast it could extend them in front of its body and use them as an additional pair of limbs. The Greeks believed that if captured the onocentaur would refuse all food and allow itself to die from starvation.

The upper, human part of the onocentaur was highly rational, but the lower part of his body was wild and dissolute. Being half-human and half-beast it symbolized hypocrisy and unreliability, like the person who talks of doing good, but whose actions in fact are evil.

SIREN

Sirens were originally beings with the head of a woman and the body of a bird. There were three of them, and they sat, surrounded by the bones of their victims, in a meadow overlooking the sea, from where their sweet music enchanted sailors, lulling them into a deep slumber and drawing them close enough to the shore for the sirens to seize and devour them. In the *Odyssey,* in which the Greek poet Homer tells the story of Odysseus's journey home from the Trojan War, Circe, the enchantress, warns of them. No human can resist their music, she says, so as their ship approaches Odysseus must soften beeswax and plug his sailors' ears with it. If he wishes to listen to their music he must order his men to bind him hand and foot and lash him securely to the ship's mast.

Early in the 12th century, however, the Norman poet Philippe de Thaon compiled a bestiary in which he described sirens as having both the feet of a bird and the lower body of a fish. This led to some confusion among the authors of later bestiaries. Most sirens appear in the bestiaries with the lower body of a fish, so they resemble a mermaid, but in addition they often have wings on their backs or on waists, and sometimes they have bird claws as well as a fish tail. There was no doubt at all about their nature, however, and the bestiary writers described their behavior very much as Homer had done.

The message was also clear. Those who delight in the pomp and vanity of the world and spend their time listening to comedies, tragedies, and music—in modern terms watching too much TV and listening to pop music—will lose their mental vigor. Then they will fall prey to their enemies, who will destroy them.

There was also another siren, which had nothing to do with the sea. It was a white serpent that lived in Arabia. It could move faster than a horse and according to some reports it could also fly. This siren was venomous, with a poison so powerful that if it bit someone that person would be dead before feeling the pain of the bite.

SATYR

The satyr was one of five types of ape. It was pleasing to look at and constantly restless. Sometimes the bestiaries depicted it with horns and cloven hoofs, showing that the authors had taken their descriptions from classical sources. In Greek mythology the satyrs were companions of the gods Pan and Dionysus that roamed in the mountains and forests. They had large, pointed ears, long, curly hair, and full beards. They were cowardly and evasive, but also dishonest and dangerous to know. The Romans later adopted them, and that is when they acquired their goatlike appearance, with horns and cloven hoofs.

Pan, the Greek god of green fields, was one of the most famous satyrs. He was the son of Hermes, the messenger of the gods, and a nymph. He loved music and dancing and often played on the pan pipes that he had invented. In the fifth century B.C.E. a new type of drama developed in Athens in which a chorus of satyrs and sileni—older satyrs—commented on the main action of the play and ridiculed the mythological events it portrayed. This is the origin of satire.

Satyrs were associated with drunkenness and every kind of debauchery. Consequently, they had close links with the devil and sought to lead people into sin.

PHOENIX

Possibly the most famous of all mythological birds, the phoenix symbolized the death and Resurrection of Christ. In the Middle Ages people believed the bird really existed and the bestiaries cite its life cycle as evidence that the Resurrection really happened. Birds have been created to teach people, the bestiaries argue, and the phoenix displays all the signs of the Resurrection.

There was only ever one phoenix. In one account it had a crest on its head like that of a peacock, its throat and breast were red, its body gleamed like gold, and toward the tail its feathers were sky-blue. According to some versions of its story it lived in India, others held that it lived in Arabia.

All the versions agree that the phoenix lived for 500 years, and when it felt itself growing old it prepared for its own death. There are several versions of what happened then. Some bestiaries asserted that the aging bird filled its wings with sweet-smelling spices and flew to Heliopolis in Egypt, where a priest prepared an altar with firewood piled on top. Other versions of the story say that the phoenix collected its own firewood and built its own pyre, adding frankincense, myrrh, and other aromatic substances. The phoenix sat on top of the pyre and struck its beak against a stone to cause a spark that ignited the wood. Alternatively, it spread its wings to the Sun, until they were hot enough to ignite the fire. Once the fire was alight the phoenix fanned it with its wings and was soon burned to ashes.

The following day a small, wormlike animal appeared among the ashes. After two days the worm was becoming more birdlike. On the third day the fully restored phoenix rose from the ashes, saluted the priest, and flew away, back to its Indian or Arabian home.

In another version of the story, when it was time for the phoenix to die it built a chrysalis of frankincense, myrrh, and other sweet substances that might remove the stench of death, crawled into the chrysalis, and died there. The worm emerged from the chrysalis the following day, and on the third day the phoenix was restored.

The resurrection of the phoenix reflected Christ's Resurrection, and the chrysalis had its own symbolism. It demonstrated, in the words of one bestiary, "that the Author and Creator of birds does not suffer his saints to die eternally, but wishes to restore them by using

His own life-force." Those reading or hearing the story were urged to renounce their sinful ways and let Christ become their chrysalis of faith and their sheath to shelter them.

AMPHISBAENA

Bestiaries included many serpents, but not all of them were snake-like. Many possessed ears and legs and some had wings as well. The amphisbaena was the most extraordinary of them all, however, because not only did it possess ears, legs, claws, and wings, it also had two heads, one at the end of its neck and the other at the tip of its tail. Both were venomous. Its eyes glowed.

The amphisbaena was used widely in medieval illustrations and was often shown, like the one below, with the head on its neck seizing the tail just behind the second head. How the animal moved puzzled the bestiary authors—as well it might! Its name is derived from two Greek words, *amphis*, meaning "both ways," and *bainein*, which means "to go." This implies that it could move in either direction; it is difficult to say what "forward" and "backward" mean for an animal with a head at each end. Some suggested it might form itself into a hoop and roll along, others that the hoop it formed was horizontal rather than vertical and that it scuttled along in any direction. The animal symbolized anything that was ungodly or unnatural.

In the Middle Ages, the amphisbaena acquired uses in folk medicine. Wearing the skin of an amphisbaena cured colds and arthritis, for instance, and eating the meat of

Despite having legs with claws and wings, the amphisbaena was a serpent—one with two heads. This illustration is from the bestiary held at Aberdeen University, Scotland. *(University of Aberdeen)*

one attracted lovers. Many of these beliefs originated in lands around the Mediterranean and may have been based on worm lizards, about 130 species belonging to the family Amphisbaenidae, found around the Mediterranean, in Africa, and in Central and South America. These are small reptiles, most less than 6 inches (15 cm) long, which burrow below the ground. They resemble worms, skin covering their eyes and ears. They are not venomous, but they are carnivores, feeding on small invertebrates.

HYDRUS

Throughout history, most people have feared snakes and regarded them as evil. The medieval bestiaries were no exception, and all but one of the bestiary serpents were in one way or another agents of the devil. That other one was the hydrus, which lived in the River Nile and symbolized Christ. The hydrus was not to be confused with the hydra of Greek mythology that grew three heads for every one that was severed, but that Hercules killed. The hydra, said the bestiaries, was just a myth, whereas the hydrus was real. Nevertheless, the two were sometimes confused and some medieval texts showed the hydrus with three heads. The hydra of the Hercules legend was said to be a many-headed water dragon that lived in the swamp of Lerna, in Arcadia. In fact, the bestiary authors wrote, the hydra was really a place where water gushed out from many points, flooding cities, and when one source was sealed another opened.

The illustration on page 92 shows a variety of fish and other aquatic animals, dominated by a splendid sea horse. It is taken from the opening page of the section on "the swimmers" in a 13th-century bestiary and *herbarium*—a book of dried plant specimens—and the hydrus is shown in the bottom left corner. It has the tail of a fish, one pair of legs, and horns.

The hydrus was the enemy of the crocodile (see "Crocodile" on pages73–75). It was a water snake, *hydros* being the Greek word for "water," and it was dangerous, though not lethal. Anyone bitten by it was likely to swell up with what was called ox-sickness because it could be cured by smearing cattle dung on the affected part. The hydrus looked out for a crocodile sleeping on the riverbank with its mouth open. When it found one, the hydrus rolled itself in river mud, so its body was very slippery. Then it slid into the crocodile's mouth,

The horned animal in the bottom left corner of this illustration is the hydrus. The picture as a whole is taken from a bestiary and herbarium (book about plants) compiled in the 13th century and was probably used at the beginning of the section on "the swimmers." Among the other mythical beasts, the picture includes a very impressive sea horse. *(The British Library/Topham-HIP/The Image Works)*

which startled the crocodile and made it swallow involuntarily. Once inside the crocodile the hydrus set to work devouring its intestines, then splitting open the belly to escape, killing the crocodile in the process. Pictures of the hydrus usually show it with its head protruding from the crocodile's side and its tail extending from the crocodile's mouth. There is also an Icelandic bestiary that depicts the hydrus as a bird, and some bestiaries show it as a snake.

Death and hell, said the bestiaries, are like crocodiles, and Christ is their enemy, for Christ took on human form and descended into hell, where he devoured hell's intestines and emerged, leading out all those who had been unjustly imprisoned there.

CALADRIUS

The caladrius, sometimes called the charadrius, was a river bird. It was entirely white, without the smallest mark in any other color, and it was kept in the courts of kings. Looking directly at it could cure jaundice, and its dung was believed to cure blindness. The bird's most remarkable ability, however, was to tell whether or not a sick person would recover.

The caladrius was placed beside the sick person. If that person was fated to die, the bird would turn its head away the instant it saw the patient. If the patient was to live, the caladrius would look him or her in the face and take all the sickness upon itself. The bird would then fly high into the sky, burn off the sickness, and scatter it into the wind. The patient would immediately recover. Most illustrations of the caladrius show it looking at the patient, although sometimes two birds are shown, one looking at the patient and the other averting its gaze from the same patient, or a room with two beds, each with a patient and a caladrius.

When it turned its face away from those who were to die the caladrius represented Christ, who turned his face away from unbelievers. The bird also represented Christ when it took the sickness upon itself, healing the sick person, and it was pure white, just as Christ was pure white, without a speck of sin.

There was also another symbolism. The caladrius had a long neck and sought its food below the ground. It was classed as unclean and eating or imitating it was forbidden. In this the bird represented the contemplative person who studied religion and appeared religious, but whose life was taken up with earthly things that others should not imitate.

Fantastic Humans

Travelers returning from distant lands brought tales of strange plants and animals. They also brought stories about humans, or beings resembling humans in certain respects, who were unlike any that their listeners had ever seen or could imagine. In his *Etymologiae,* an encyclopedia in which he aimed to record all the learning of ancient and modern Europe, Isidore, the archbishop of Seville at the beginning of the seventh century, described these beings as "monstrations," or warnings of the divine will. Scholars have long debated exactly what he meant. What were these beings, what warnings were they intended to convey, and to whom were the warnings addressed?

Stories tend to improve with repeated telling, exaggerating details that make them more interesting and especially those that make them more sensational. The most likely explanation for the tales of fantastic humans begins with the way people viewed the world beyond their known borders in the days when few ventured farther than the next village. Today, sci-fi movies and TV shows, improving all the time with each advance in the quality of their special effects, depict the strange beings that inhabit other worlds. Those worlds are imaginary, but for the duration of the story the audience accepts them as real, and no one can be quite sure that such beings, or beings still more fantastical, are not to be found on planets millions of light-years from Earth. They are "the other," the beings that are different from us, that are quite literally alien. The medieval fantastic beings were the aliens of their day. Now we know that no matter where they live on Earth, people are alike in every important respect, so the truly alien has to be located in a distant world. In the ancient and medieval world, when entire continents remained unexplored, the aliens were located in distant lands, and they were the subject of adventure stories and cautionary tales.

These stories are very ancient, their true origins lost in the mists of Indo-European history. Some were firmly established in ancient Egypt and taken up and adapted by the Greeks, but the stories occur throughout Eurasia. This short chapter examines just a few of those fantastic beings. It begins with the most famous of them all—the men with the heads of dogs. Then it tells of a few of the other monstrous beings that were alleged to inhabit the remote corners of the world.

DOG-HEADED MEN

Wolves tend to avoid people. They fear them, and approach only when they are desperate for food. They will take sheep and deer, of course, so they compete with humans for food. For centuries, shepherds have needed to guard their flocks from them, usually with the help of domestic dogs bred for the purpose. People were afraid of wolves because, although unprovoked attacks are rare, a wolf will defend what it regards as its food against any human who tries to rob it.

In the Middle Ages wolves symbolized the wild. Domestic dogs, in contrast, were familiar and welcome as companions and servants. Many dog lovers describe dogs as being almost human, and throughout history they have remained close to human society. In addition to dogs and the wild wolves, there developed a third category of being, the man with the head of a dog. The dog-headed man was more human than a dog and his social behavior was like that of a human. Dog-headed men lived in faraway lands on the edge of the world, in particular India and Ethiopia. That may be because these and other fantastic humans occur in Indian myths and also because in Europe the most sought-after breeds of dog were imported from Egypt—origin of the greyhound—and Asia. Ctesias, a Greek physician and historian who lived in the fifth century B.C.E., believed they lived in India and described them in the following words.

On these mountains there live men with the head of a dog, whose clothing is the skin of wild beasts. They speak no language, but bark like dogs, and in this manner make themselves understood by each other. Their teeth are larger than those of dogs, their nails like those of these animals, but longer and rounder. They inhabit the mountains as far as the River Indos. Their complexion is swarthy. They are extremely just, like the rest of the Indians with whom they associate. They understand the Indian language but are unable to converse, only barking or making signs with their hands and fingers by way of reply, like the deaf and dumb. . . . They live on raw meat and number about 120,000.

As European merchants and explorers visited more and more regions of the world without finding living examples of them, however, their territory grew progressively smaller. Marco Polo, the 13th-century Venetian merchant who traveled through much of Asia, including India (see "Marco Polo and the Gardens of Kinsai" on pages 29–34), located them in the Andaman Islands, in the Indian Ocean. He described them in the following excerpt from his *Travels*:

Andaman is a very big island. The people have no king. They are idolaters and live like wild beasts. Now let me tell you of a race of men well worth describing in our book. You may take it for a fact that all the men of this island have heads like dogs, and teeth and eyes like dogs; for I assure you that the whole aspect of their faces is that of big mastiffs. They are a very cruel race: whenever they can get hold of a man who is not one of their kind, they devour him. They have abundance of spices of every kind. Their food is rice and milk, and every sort of flesh. They also have coconuts, apples of paradise [*Malus pumila*, the cultivated apple], and many other fruit different from ours.

The illustration on page 97, painted in about 1400, shows a group of dog-headed men in the Andaman Islands, discussing and perhaps buying and selling fruit. They wear fashionable European clothes and appear to live in a city with magnificent buildings.

Dog-headed men, often called by their Greek name *cynocephali*, are also based in part on the baboon, which has a doglike head. In ancient Egypt the god Babi was a hamadryas baboon (*Papio hamadryas*), and Hapy, the son of Horus, was depicted with the body of a man and the head of a dog (or possibly jackal). Cynocephali also

appeared in Christian art. St. Christopher is sometimes represented as a dog-headed man in icons of the Eastern Orthodox Church. One of the many Christian legends tells how Saints Andrew and Bartholomew traveled to Parthia, a land to the south of the Caspian Sea, where they encountered Abominable, who was a citizen of a city of cannibals and had the face of a dog. The saints converted him, and released him from his doglike nature.

The dog-headed men who were believed to inhabit the Andaman Islands in the Indian Ocean, depicted in a painting created around 1400, held at the Bibliothèque nationale in Paris, France *(Archives Charmet)*

The general Christian view was that the dog-headed men were sufficiently similar to humans to be suitable for conversion. Although some were savage prior to their conversion, all of them practiced agriculture and lived in well-ordered societies.

BLEMMYES, CYCLOPS, SCIAPODAE, AND PANOTII

Ctesias may have derived many of his stories from Indian legends of strange peoples, among whom the dog-headed men were far from being the strangest. Pliny the Elder (see "Pliny, Collector of Information" on pages 21–26) said of the Blemmyes: "It is said that the Blemmyes have no heads, and that their mouth and eyes are put in their chests." Shakespeare also knew of them. In *Othello, The Moor of Venice* (1.3), Othello described his travels and adventures, mentioning:

> ". . . the Cannibals that each other eat,
> The Anthropophagi, and men whose heads
> Do grow beneath their shoulders."

What can such a creature look like? The illustration below shows one among a group of other fantastic men, including a dog-headed man. It is taken from the *Miroir historial*, written in about 1332; it is a French translation by Jean de Vignay of the *Speculum Historiale*, written much earlier in Latin by Vincent de Beauvais (died 1264).

Blemmyes had no heads, but their faces were in the middle of their chests. Here a blemmye (and a portion of a second one on the right) appears to be discussing some matter with a dog-headed man, a wild man eating an eel, and a member of the Sciritae, an Indian tribe with nostrils like those of a snake instead of a nose. *(J. Paul Getty Museum)*

The Blemmyes were said to live in Africa, somewhere to the south of Egypt, and the legends about them were based on a real people. The real Blemmyes were a nomadic tribe living in Nubia, in what is now Sudan. Their military and political power grew during the second century C.E., and they fought the Romans several times, a conflict that ended with the total destruction of the Blemmyan army. No one knows the origin of the idea that their faces were on their chests, although some authorities have suggested they may have had a habit of hunching their shoulders very high.

The Cyclops (plural Cyclopes) was a giant with a single eye in the middle of his forehead. In Greek mythology there were two generations of Cyclopes. The first were metalworkers and blacksmiths who worked for the gods. The second generation, which was probably added much later, were lawless, violent monsters.

Some scholars have suggested that stories about the Cyclopes might have originated with the discovery of the skulls of *Deinotherium giganteum*—a species of elephants that arrived on several Mediterranean islands about 2 million years ago. As often happens to large animals that colonize islands, the elephants became smaller in subsequent generations. The resulting dwarf elephants, with skulls about double the size of a human skull, became extinct about 8,000 years ago. An elephant skull has a central opening for the nose, which is elongated into a trunk, but people who had never seen an elephant could not have known about the trunk and might have supposed that the single opening was an eye socket. Other scholars think the stories may have referred to the habit of Greek blacksmiths of wearing an eyepatch to protect one eye from sparks and cinders.

As an example of the way stories improve as they pass from one storyteller to another, Pliny the Elder, in his *Natural History* written in the first century C.E., repeats what he read in the works of Ctesias, written in the fifth century B.C.E. Ctesias, wrote Pliny, "describes a tribe of men called Monocoli who have only one leg, and who move in jumps with surprising speed; the same are called the Sciapodae (Shadow-Foots) tribe, because in the hotter weather they lie on their backs on the ground and protect themselves with the shadow of their feet."

The Sciapodae lived either in Ethiopia or in India, and images of them were popular in the Middle Ages. There are carvings of Shadow-Foots in some English churches dating from that time and an engraving of one in the *Liber Chronicarum* (Nuremberg Chronicle), a history of the world from its creation until 1493, the year the book was published. Some scholars have suggested that the legend may be based on Indian yogic practices that involve meditating while standing on one foot.

A Panotian is also shown in the *Liber Chronicarum*. The Panotii lived in the far north of Europe, where the climate is very cold. They had adapted to the conditions by acquiring immensely large ears that they wrapped around themselves to keep warm and snug at night. Both Ctesias and Pliny described them and also a similar tribe called the Pandai that lived in India. The Panotii could also use their ears as wings, to fly away from danger.

When Europeans first began exploring the Americas, one of their objectives was to discover fantastic humans like those of European legend. Several contemporary accounts, including one by Columbus, mention tribes of Cynocephali, at least some of which were cannibalistic. The Spanish painter Diego Velázquez (1599–1660) is said to have asked Hernán Cortés (1485–1547) to search for the Panotii and in his book *Relation of a Voyage in Guiana,* published in 1613, Robert Harcourt (ca. 1575–1631) claimed that a Panotii tribe was living beside the River Maroni in Surinam. Eventually, though, as Europeans explored even the most remote regions of the world, the tales of fantastic people died away.

Observation and Understanding

Until the rise of the modern pharmaceutical industry in the 19th century, most medicines were obtained from herbs. Physicians were taught botany at medical school, because they needed to be able to identify medicinal plants. Consequently, the study of plants advanced much more quickly than the study of animals. Wild animals had no obvious use, and domestic animals had been tamed in prehistoric times and improvements to their productivity were achieved through long-established methods of selective breeding. Throughout the Middle Ages, therefore, there was little challenge to the fantastic but highly popular descriptions of animals to be found in the bestiaries.

By the 16th century, however, animals were attracting the interest of naturalists, and an alternative approach to the study of them was starting to emerge. Naturalists were beginning to base their accounts of animals on firsthand observation rather than tales that had been handed down over the centuries and never questioned. Scholars began to study animal anatomy and to make anatomical comparisons between species—a task that led them to seek definitions for the concept of species. This chapter describes a few of the early steps taken in the new direction and tells of the work of the naturalists involved.

CONRAD GESSNER, THE SWISS PLINY

Natural History, the vast encyclopedia compiled by Pliny the Elder (see "Pliny, Collector of Information" on pages 21–26) in the first century C.E., dominated the study of animals throughout the medieval period. The authors of bestiaries made extensive use of it, and by the late Middle Ages it had been so well established for so long that it was difficult to challenge its descriptions.

The first sign of change came between 1551 and 1558 with the publication in Zürich of a new four-volume illustrated encyclopedia by Conrad Gessner (1516–65), *Historia Animalium* (Natural history of animals). Quadrupeds (meaning mammals), birds, fishes, and snakes each had a volume devoted to them. This encyclopedia aimed to bring together all the knowledge that had accumulated since classical times, and many of its descriptions were based on direct observation. Historians of science regard *Historia Animalium* as the starting point of modern zoology. Apart from Pliny's *Natural History,* it was the first comprehensive work on the subject since Aristotle's *Historia Animalium,* written in about 343 B.C.E. Parts of the work were translated into the English language in 1608, and the full English title gives a good idea of the contents: *The Historie of Foure-Footed beastes. Describing true and lively figures of every Beast, with a discourse of their several Names, Conditions, Kindes, Vertues (both naturall and medicinall), Counties of their breed, their love and hate for Mankinde, and the wonderfull worke of God in their Creation, Preservation, and Destruction. Necessary for all Divines and Students because the Story of every Beaste is amplified with Narratives out of Scriptures, Fathers, Phylosophers, Physitians, and Poets; wherein are declared divers Hyeroglyphicks, Emblems, Epigrams, and other good Histories.*

The work included some of the fantastic animals found in earlier works and in the bestiaries, but Gessner qualified those descriptions by stating that there was no good evidence for their existence. The animals were not classified, however, beyond being placed in the four broad groups corresponding to the four volumes. Instead they were arranged alphabetically within each volume, so animals were placed together if they were of the same general type. The concept of the species had not yet been clarified, and scholars continued to believe, as had the medieval bestiary writers, that fantastic animals could result from the mating of different species. They believed that each animal had essential characteristics that were fixed at the time of its creation and that these were maintained through the mating of like with like, but when two different kinds of animals mated the offspring would possess some combination of characteristics derived from each of them. Also, scholars still believed in *spontaneous generation*—the idea that living animals could emerge spontaneously from putrefying matter. These two beliefs made it impossible to define species clearly and to classify organisms on the basis of relationships arising from common ancestry.

Conrad Gessner (his first name is sometimes spelled Konrad and his family name Gesner, but Conrad Gessner is how he himself spelled it) has been called the German Pliny, but he was Swiss, so the Swiss Pliny would be more accurate. He was born in Zürich on March 26, 1516, the son of Urs Gessner, a furrier, and Agathe Fritz. He began his education in Zürich at the Fraumünster School and studied theology at the Collegium Carolinum (which later became the University of Zürich), but wars triggered by the Reformation were being fought through the Swiss cantons during his childhood and in 1531, when he was 15, his father died in the battle of Kappel. The family had not been wealthy, but his teachers helped the young Gessner to continue his studies at the universities of Strasbourg and Bourges, and he taught Greek at Strasbourg in 1532. With further help from his patrons, Gessner was also able to study Latin and Greek literature, rhetoric, and natural and moral philosophy in Basel and Paris. Protestants were being persecuted in Paris, however, and in 1534 he was forced to return to Strasbourg and then to Zürich. While in Zürich he married Barbara Singerin in 1536. The marriage was unhappy, and to support his wife Gessner had to teach in an elementary school for a time, before further help from his patrons allowed him to return to Basel to study medicine.

Gessner compiled a Greek–Latin dictionary in 1539, and this led to his appointment as professor of Greek at the Academy of Lausanne, where he also continued his studies. In 1540 he moved to the medical school at Montpellier, France, where he met Pierre Belon and Guillaume Rondelet, two of the leading natural historians of their generation. Gessner received his degree in medicine at Basel in 1541. He then returned to the Collegium Carolinum in Zürich where he lectured on mathematics, physics, astronomy, philosophy, and ethics. He also practiced medicine, and it was around then that he began writing and publishing work on a wide range of subjects. In 1545 he published his *Bibliotheca Universalis* (Universal bibliography) in which he tried to list, in Latin, Greek, and Hebrew, every author who had ever lived with the titles of all their works; Gessner is also known as the father of bibliography. In 1555 he published *Mithridates de Differentis Linguis* describing about 130 languages and with the Lord's Prayer in 22 languages. As well as his zoological encyclopedia, Gessner also wrote about plants.

In 1564 the Holy Roman Emperor Ferdinand I (1503–64) ennobled Gessner, who became von Gessner, with a coat of arms. Plague broke out in Zürich in 1565. Gessner died on December 13, while ministering to victims of the disease.

PIERRE BELON, AND THE SKELETONS OF BIRDS AND MAMMALS

Although he made a major contribution to the emergence of zoology as a science, languages and literature were Conrad Gessner's first loves and in his animal encyclopedia he was concerned primarily with collecting together everything that was already known. He sought to expand on the work of Aristotle, Pliny, and other Greek and Roman authors rather than breaking new ground. In 1555, however, a book was published in Paris that really did offer something new and different. *L'Histoire de la Nature des Oyseaux* (Natural history of birds) by Pierre Belon (ca. 1517–64) described about 200 birds, but rather than arranging them in alphabetical order as Gessner had done, it placed them in groups of birds that were similar to each other. In addition, its 163 hand-colored illustrations were lifelike, showing the birds as they appear in their natural surroundings, much as they might be shown in a modern field guide. Most important of all, the book described the anatomy of the birds.

Belon's book contained one especially interesting illustration. This showed the skeleton of a human beside the skeleton of a bird, but with the bird standing in an upright position and the same size as the human. The diagram labeled the major bones of the human and bird skeletons with letters, using the same letters for each skeleton

---

(opposite) A human arm, a seal's flipper, a horse's foreleg, and a bird's wing are all based on the same set of bones. In this drawing the bones of the human arm are named, and the colors indicate the same bones—with the same names—in the limbs of the other animals. The bones are modified in each animal, but the limbs are constructed in the same way. These limbs are said to be homologous structures.

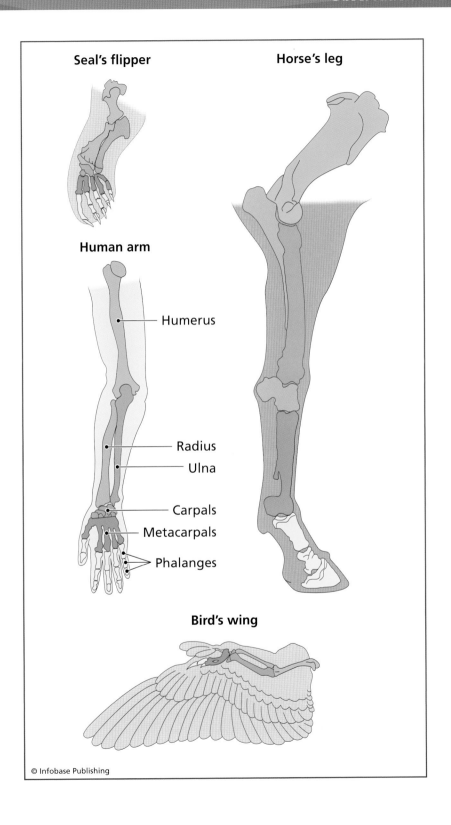

Seal's flipper

Horse's leg

Human arm

Humerus

Radius

Ulna

Carpals

Metacarpals

Phalanges

Bird's wing

so the reader could easily see the correspondences between bones in the two. This was the beginning of *comparative anatomy*—the study of the anatomical differences and similarities between different species. Comparative anatomy later showed that all vertebrates share the same basic skeletal plan. Where two species share a similar structure these are said to be homologous. For example, the human arm, horse's foreleg, seal and whale flipper, and bird and bat wings all contain the same set of bones: *humerus, radius, ulna, carpals, metacarpals,* and *phalanges.* These vary in their proportions and in the horse the radius and ulna are fused into a single bone, but the illustration on page 105 shows their clear similarities.

Animal *taxonomy*—the scientific classification of organisms—identifies relationships on the basis of homologous structures, taking

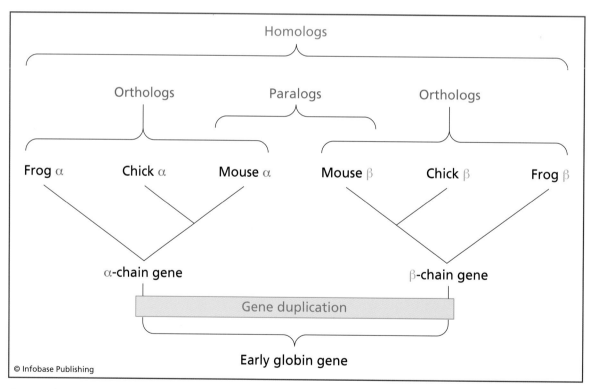

© Infobase Publishing

Genes inherited by different species from a common ancestor are also homologous. In this case, however, gene duplication may produce gene copies with different functions. Consequently, there are two types of homologous genes. Genes in different species that are inherited from a common ancestor are said to be orthologs. Genes in a single species that result from gene duplication are said to be paralogs.

HOMOLOGY, ANALOGY, ORTHOLOGY, AND PARALOGY

The arm of a human is very different in appearance from a seal's flipper, a whale's flipper, or the wing of a bat or a bird. Yet the bones that give each of these animal limbs their structure are very similar. The relative sizes of the bones vary, but not the order in which they are arranged in the limb. These similarities strongly suggest that birds and mammals have inherited their skeletal structures from an ancestor common to them all. The structures—the arm, flipper, and wings—are said to be homologous. Homology is accepted as scientific evidence of descent from a common ancestor.

Most insects also possess wings, and like the wings of birds and bats they are used for powered flight. Insect wings have an entirely different construction, however. They serve a similar purpose, but the insect has not inherited its wings from an ancestor it shares with mammals or birds. These structures—insect wings and the wings of birds and bats—are said to be analogous. Analogy is not evidence for descent from a common ancestor.

Organisms also inherit homologous genes, but genes may undergo duplication, producing a pair of genes that may have different functions. There are, therefore, two types of homologous genes. Identical genes that different species inherit from a shared ancestor are said to be orthologous, or orthologs. Two versions of a gene, present in a single organism, that result from the duplication of a gene inherited from an ancestral species are said to be paralogous, or paralogs.

care to distinguish them from *analogous* structures, which perform similar functions but are structurally dissimilar. The wings of an insect and a bird are both used for flight, but they are structurally different and were not inherited from a common ancestor of both birds and insects (see the sidebar "Homology, Analogy, Orthology, and Paralogy" above).

Species inherit genes from ancestral species, so the identical genes that code for the same protein in different species are also homologous. In the case of genes, however, there are two types of homolog, arising from the fact that when cells divide genes can sometimes be duplicated. The illustration on page 106 shows the difference between these two types of homolog.

Pierre Belon was born near Le Mans, a town southwest of Paris, France. He studied medicine in Paris and then moved to Wittenberg, Germany, where he studied botany. On his return to Paris, Belon found a patron in François de Tournon (1489–1562), a wealthy diplomat and cardinal, who supplied him with funds to undertake a journey of scientific discovery through Greece, the Near East, Egypt, Arabia, and Palestine. Belon embarked on his travels in 1546 and returned in 1549. He published an account of his experiences in 1553, with a revised edition in 1555. Belon also wrote *Histoire Naturelle des Étranges Poissons* (Natural history of exotic fish) in 1551. One evening in April 1564, Belon was murdered while passing through the Bois de Boulogne, a park in Paris.

GUILLAUME RONDELET AND THE STUDY OF FISHES

Pierre Belon was not the only French naturalist studying and writing about fish in the mid-16th century, and, although Belon's work was published a few years earlier, Guillaume Rondelet (1507–66) is usually regarded as the founder of *ichthyology*, with his *Libri de Piscibus Marinus* (Book of marine fish) published in 1554 and *Universae aquatilium Historiae* (Universal aquatic natural history) published in 1555. Rondelet wrote in Latin, but in 1558 the two titles appeared in a single volume, translated into French as *L'Histoire Entière des Poissons* (Complete natural history of fishes).

Rondelet's book contained 490 illustrations. These included accurate drawings of their subjects, but also anatomical drawings based on dissections Rondelet had performed. In some cases several illustrations were devoted to an individual species. The text was based on observations of the fishes in their natural surroundings, with outlines of what would now be called their ecology, and for each species it described the general appearance, anatomy, physiology, movement, and reproduction. Unlike most modern books on ichthyology, Rondelet's work included, where appropriate, details of the therapeutic uses of the animals and recipes for cooking them.

At that time the word "fish" was taken to mean any aquatic animal. Consequently, Rondelet included whales and other aquatic mammals and also many invertebrates. The work described a total of 440 species, of which 240 were fish in the modern sense.

Guillaume Rondelet was born on September 27, 1507, at Montpellier, in southern France, which is where he began his education. He then studied medicine in Paris before returning to Montpellier, where he qualified as a physician in 1537 and became a professor in the medical faculty of the University of Montpellier; he was sometimes known by the latinized version of his name, Rondeletius. One of his students was Michel de Nostredame (1503–66), better known as Nostradamus, the apothecary and self-styled prophet and astrologer. Rondelet made several journeys through France, the Netherlands, and Italy. As well as his work with fishes, Rondelet wrote several works on medicine and pharmacology. He died from dysentery at Montpellier on July 30, 1566.

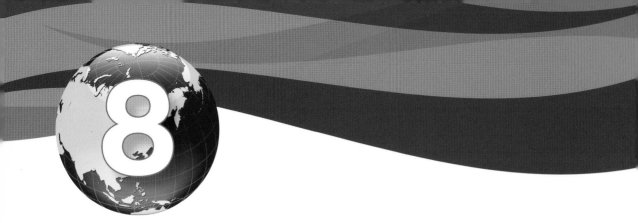

Classifying and Experimenting

By the beginning of the 17th century, naturalists were in possession of detailed and fairly accurate descriptions of large numbers of animals. They had moved beyond the inaccuracies of the medieval bestiaries and required authentic specimens, or at least trustworthy descriptions, before they would accept previously unknown animals as real. The manticoras, dragons, basilisks, and other fantastic beasts had been banished to the realms of mythology, where they have remained ever since.

As yet, however, naturalists had no acceptable, workable system for classifying the animals they studied. Every animal had a common name, of course, but this name varied from place to place and the same name might be applied to more than one animal. This problem continues to the present day. The animal Americans call the caribou, for instance, is known in Europe as the reindeer; the American moose is called an elk in Europe; and Americans know the European red deer as the elk. That is not all. The puma, cougar, mountain lion, and panther are all the same animal. It is very confusing.

Solving the problem was far from simple, but serious attempts to do so began in the 17th century, and the basis of the classification biologists use today was in place by the middle of the 18th. This chapter tells how the science of taxonomy developed.

At the same time, the invention of the microscope allowed naturalists to study animals in much closer detail, and the chapter describes the first microscopes and how they were used. For the first time, scientists were beginning to construct theories about the way animal bodies work. It was during this period that the idea arose of animals being machines.

JOHN RAY AND FRANCIS WILLUGHBY

Pierre Belon and Guillaume Rondelet had studied the anatomy of the animals they drew and described. They recognized that a simple description and depiction of the external appearance of an animal was not enough to identify it positively, but they still had no general system for arranging animals in an order based on their physical structure. The first naturalists to make a serious attempt at devising such a system were John Ray (1627–1705) and his pupil and close friend Francis Willughby (1635–72). Ray wrote mainly about plants and Willughby about animals, but after Willughby's death Ray completed much of his work, publishing it under Willughby's name, although he had written most of it himself. Ray also wrote *Synopsis Methodica Animalium Quadrupedum et Serpentini Generis* (Methodical synopsis of the kinds of quadrupeds [mammals] and reptiles), published in 1693, *Historia Insectorum* (Natural history of insects), published posthumously in 1710, and *Synopsis Methodica Avium et Piscium* (Methodical synopsis of birds and fishes), published in 1713. Ray based his classification on the whole of the evidence before him, rather than concentrating on a particular aspect and ignoring others.

John Ray was born on November 29, 1627, in the village of Black Notley, near Braintree, in Essex. His father was a blacksmith and his mother was a healer and herbalist who may have imparted her interest in plants and natural history to her son. John attended school in Braintree and entered the University of Cambridge in 1644 to study languages, mathematics, and natural science. He spent nearly two years at Catherine Hall (now St. Catherine's College) and then moved to Trinity College. He graduated in 1648 and became a fellow of Trinity College in 1649. In 1651 he was made a lecturer in Greek, in mathematics in 1653, and in humanities in 1655, and he became a junior dean in 1658. In 1650 Ray suffered a serious illness and while he was recovering he spent many hours walking through the countryside around Cambridge. These walks revived his interest in natural history, which became his principal occupation thereafter.

Ray was ordained a minister in 1660, the year that Charles II was restored to the British throne and religious practice began to change after the years of Puritan dominance. In 1661 Parliament passed the Act of Uniformity. This required all clergy to sign a declaration of assent that they would use only the rites and ceremonies set down in the Book of Common Prayer. Ray felt unable to do so, apparently because he considered it an unlawful oath. It meant he had to leave the church, and, since membership of the university was open only to members of the Church of England, in 1662 Ray had to resign, bringing his academic career to an end. He was elected a fellow of the Royal Society in 1667.

Ray had lost his only source of income and for the rest of his life he depended on the financial support of his friends, the foremost of whom was his former student Francis Willughby, who was independently wealthy. In 1663, accompanied by two students, the two men set off on a tour of Europe studying the flora and fauna and collecting specimens. Ray returned to England in 1666 and went to live in Willughby's home in Warwickshire, while Willughby explored Spain. When Willughby returned the two collaborated in writing the results of their researches, but in 1672 Willughby died from pleurisy, leaving Ray a small annual sum of money to educate his two sons. In 1673 Ray married Margaret Oakley, a governess in the Willughby household. The couple left the Willughby home in 1676, moving first to Sutton Coldfield, Warwickshire, in 1677 to Falborne Hall, Essex, and in 1679 to Black Notley, where they remained. Ray spent his time writing books and letters to scholars all over Europe, but his health was poor. He died at Black Notley on January 17, 1705.

Francis Willughby was born on November 22, 1635, at Middleton Hall, Warwickshire, the son of Sir Francis Willughby and Cassandra Ridgway. He was educated at Bishop Vesey's Grammar School in Sutton Coldfield and at Trinity College, Cambridge. He married Emma Barnard in 1667. They had two sons and one daughter. Willughby became an authority on birds and fish. He died on July 3, 1672.

THE INVENTION OF THE MICROSCOPE

While naturalists such as Ray and Willughby studied the large-scale structure of organisms in order to identify and classify them, other researchers were examining the finer details using powerful lenses.

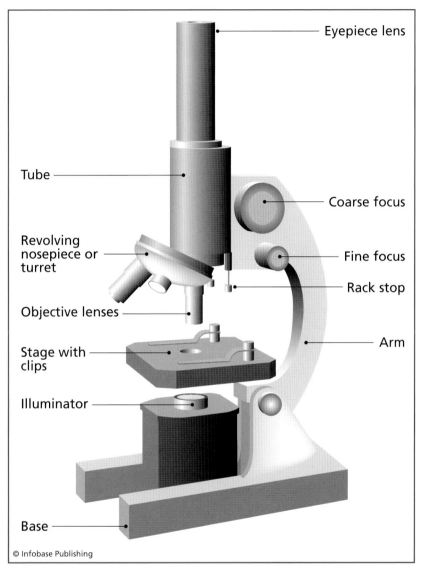

Eyepiece lens

Tube

Coarse focus

Revolving
nosepiece or
turret

Fine focus

Rack stop

Objective lenses

Stage with
clips

Arm

Illuminator

Base

© Infobase Publishing

The simplest modern light microscope is a sophisticated instrument. The
eyepiece lens (usual magnification 10× or 15×) is mounted at one end
of the tube. At the other end the revolving turret carries two or more
objective lenses (usual magnifications 4×, 10×, 40×, and 100×). The total
magnification is that of the objective lens multiplied by the eyepiece lens.
The arm supports the tube. The slide fits on the stage; clips prevent it
moving about. The illuminator is a light that shines upward through the
slide; some microscopes use a mirror to reflect ordinary light. The rack stop
prevents the objective lenses from touching the stage.

Every modern biology student learns to use a microscope. The instrument is so familiar it is easy to overlook just how sophisticated it is. Even the simplest light microscope relies on lenses that have been ground and polished to very precise specifications. The illustration on page 113 shows the structure of the instrument. It is called a light microscope because it uses light to illuminate the specimen being examined. A light microscope cannot resolve an object that is smaller than one-half the wavelength of light (0.275 μm; 0.00001 inch). The much more advanced and more powerful alternative is an electron microscope, of which there are several types.

The microscope has a long history that began in the 9th century with the invention of magnifying lenses made from colorless glass to help with reading, probably by Abbas Ibn Firnas (810–87), who lived in Córdoba, in al-Andalus, in Moorish Spain. By the 14th century the lenses that Abbas Ibn Firnas called "reading stones" were being used in spectacles. There is a fresco in a church in Treviso, Italy, that shows an elderly scholar studying a manuscript and clearly wearing spectacles. Tommaso da Modena (ca. 1325–79) painted the fresco in 1352, and it is the earliest definite evidence of the existence of spectacles. It is very likely that they were invented in Italy not many years earlier.

Spectacles were of obvious value, and their success encouraged lens makers to experiment. Those experiments led to the invention of the first simple microscope, consisting of a lens mounted at one end of a tube, with the object to be examined at the other end. No one knows who made that discovery, but in the early 1590s certain Dutch spectacle makers discovered that several lenses mounted inside a tube achieved a much higher magnification than was possible for any single lens. They had made the first compound microscope—compound because it used more than one lens. Most historians believe it was Sacharias Jansen, also spelled Zacharias Janssen (1580–1638), assisted by his father Hans, who made the first microscope. Others credit Hans

A microscope that was made in the 17th century *(Giraudon/ Art Resource, NY)*

Lippershey (1570–1619). Both Lippershey and the Janssens lived in Middelburg, in the Dutch province of Zeeland. Galileo (1564–1642) heard of their work and improved on it. He discovered why lenses magnify objects, and he made a microscope with a device for focusing. The illustration on page 114 shows a 17th-century microscope. Its eyepiece is mounted on a tube that slides inside the tube holding the objective lens, thus allowing the user to focus the microscope.

By the middle of the 17th century naturalists were using microscopes to study living tissue. In England, Robert Hooke (1635–1703), one of the most brilliant men England has ever produced (see the sidebar on page 116), improved on the existing microscope and made exquisite drawings of the images his lenses revealed. He studied a thin sliver of cork and found it was divided into countless tiny compartments that he called cells because they reminded him of monks' cells. He drew a flea and many other small insects in fine detail. Hooke published his drawings and written descriptions in 1665 in a book with the title *Micrographia, or some Physiological Descriptions of Minute Bodies, made by Magnifying Glasses, with Observations and Inquiries thereupon.* Hooke's first description was of the point of a needle. He followed this with the edge of a razor and then fine lawn linen and silk taffeta.

Many people read Hooke's *Micrographia* and were impressed by it. Among them was a Dutch draper living in Delft who used a microscope to examine the quality of woven cloth. His name was Antoni van Leeuwenhoek (1632–1723), and Hooke's book inspired him to seek higher magnifications. Van Leeuwenhoek made his own lenses and had found a way to melt glass, draw it out into a long thread, and then reheat the end of the thread to make a small sphere. Ground and polished, the spheres he made were his lenses, some giving a magnification of 270×, and he made many observations with them, carefully drawing what he saw. His examination of cloudy water from a lake revealed a new world of tiny living organisms. He saw red blood cells, human spermatozoa, bacteria, and protozoa. In one of his experiments van Leeuwenhoek scraped a little tartar from one of his teeth, placed it under his microscope, and found it contained living organisms.

In 1673, Regnier de Graaf (1641–73), a physician and anatomist living in Delft who knew van Leeuwenhoek, wrote to Henry Oldenburg (1619–77), secretary to the Royal Society of London. Oldenburg

ROBERT HOOKE: ENGLAND'S LEONARDO?

Robert Hooke (1635–1703) was one of the most ingenious experimenters who have ever lived, certainly in England and probably in the world. Some historians have rated him the equal of Leonardo da Vinci (1452–1519). Hooke was born on July 18, 1635, at Freshwater, Isle of Wight (an island off the southern coast of England). His father John was a curate—a clergyman who assists the parish priest. John Hooke wanted his son to follow him into the church, but Robert's health was poor, and he was not strong enough to sustain the long education this required. He spent much of his time alone, amusing himself by making mechanical toys. When he was older, he attended school in Oxford and, following his father's death in 1648, relatives sent him to London. At first he began an apprenticeship for which he paid the £100 he had inherited from his father, but he abandoned the apprenticeship and enrolled at Westminster School. He studied Latin and Greek, but mathematics was his strongest subject; it took him only a week to master Euclid's geometry. In 1653 Hooke entered Christ Church College at the University of Oxford under a scheme that allowed impoverished students to attend lectures free in return for performing certain menial tasks. Hooke paid his way by selling ideas for scientific instruments and improvements to existing instruments to the owners of workshops where these instruments were made. He left the university without taking a degree.

While he was at Oxford, Hooke joined a group of brilliant young scientists, one of whom was Robert Boyle (1627–91). Boyle was wealthy and paid Hooke generously for helping with his experiments. The two became lifelong friends.

In 1659 several members of this group, including Boyle and Hooke, moved to London where,

wrote to van Leeuwenhoek, initiating a correspondence that continued until van Leeuwenhoek's death. His letters—there were at least 200—were the only accounts van Leeuwenhoek ever wrote about his microscopy. The Royal Society elected him a fellow in 1680. This pleased him, but he never visited London, although many scholars visited him in Delft. After his death, his daughter Maria donated to the Royal Society a cabinet containing 26 of the microscopes her father had made. Apart from those, van Leeuwenhoek left 247 completed microscopes and 172 magnifying glasses comprising lenses mounted between metal plates.

MALPIGHI AND THE ANATOMISTS

The early microscopists, such as Hooke and van Leeuwenhoek, were investigating the possibilities of the new technology of microscopy, in 1660, they formed the scientific society that in 1662 received a charter from King Charles II and became the Royal Society of London. Hooke was appointed curator of experiments, a position that required him to demonstrate new experiments at each of the Royal Society's weekly meetings. He was paid a small salary and provided with accommodation. Oxford University awarded him a master's degree in 1663, and in the same year he was elected a fellow of the Royal Society. Hooke was secretary to the society from 1677 to 1683. The society promoted him to professor of mechanics in 1664, with a raise in salary, and in 1665 he was appointed professor of geometry at Gresham College. This was in addition to his job with the Royal Society, and he was able to combine the two salaries. He held these two posts for the rest of his life.

His work for the Royal Society involved Hooke in every branch of science and allowed him to develop his mechanical skills. It also meant, however, that he failed to finish many of his projects, and much of his work involved improving on other people's ideas. Nevertheless, his achievements were considerable. Hooke discovered that the stress placed on an elastic body is proportional to the strain it produces: This is known as Hooke's law. He invented an anchor escapement mechanism for a watch—though this led to a quarrel over priority with Christiaan Huygens (1629–95)—and he claimed to have discovered gravitation and the inverse square law before Isaac Newton (1642–1727). He improved the design of microscopes and became expert in their use. Hooke also invented the wheel barometer, in which the rise and fall of a column of mercury is translated into the movement of a needle on a dial.

Robert Hooke died in London on March 3, 1703.

but the Italian physiologist and anatomist Marcello Malpighi (1628–94) was the first person to use the microscope as a tool. Malpighi was a scientific microscopist and the studies he made were the foundation for *histology,* which is the study of biological tissues, and *embryology,* which is the study of the development of animal *embryos.*

Malpighi was born at Crevalcore, near Bologna, Italy, on March 10, 1628. He studied medicine and philosophy at the University of Bologna, where he joined a group of anatomists who performed dissections and vivisections. He qualified as a physician in 1653. In 1656 he became professor of theoretical medicine at the University of Pisà, and in 1659 he returned to Bologna as professor of theoretical medicine, later of practical medicine. He was principal professor of medicine at the University of Messina from 1662 to 1666 and then returned to Bologna, where he remained for the rest of his academic career. In 1691 Pope Innocent XII summoned him to Rome to become his personal physician. Malpighi died in Rome, in his room in the Quirinal Palace, on November 29, 1694.

Using a microscope to examine the lungs of a frog, Malpighi discovered their structure, as a vast number of tiny air sacs with a network of blood vessels running through the membranes surrounding them. This showed the way oxygen is able to enter the bloodstream by diffusing across the membrane, and it also completed the picture of the circulation of the blood. Malpighi had observed the *capillaries*—minute blood vessels with walls only one-cell-layer thick from which oxygen and nutrients diffuse into tissues and into which waste material enters for later excretion. Capillaries link the arterial system carrying blood from the heart to the venal system carrying blood toward the heart. Among many other discoveries, Malpighi found that the spinal column and nervous system consist of bundles of fibers, studied the structure of the kidney and suggested it served to produce urine, and found that bile is secreted in the liver and not the gall bladder. He also made microscopic studies of insects, especially the silkworm, and was the first to reveal that the tiniest insect possesses complex internal organs.

Malpighi studied the detailed anatomy of animals. He was an early anatomist, and the anatomists altered beliefs that had been taken for granted for centuries. It was one of the anatomists, the Dutch biological microscopist Jan Swammerdam (1637–80), who shattered the conventional view that each of the stages in the life cycle of an insect, such as the egg, larva, and mature form, was a different animal and that one somehow changed into the next. Swammerdam showed that these are all different forms of the same animal. Swammerdam had been interested in insects since childhood, but it was reading Malpighi's account of his dissection of silkworm larvae that encouraged him to use the microscope to study them. By dissecting a bee and finding the ovaries, Swammerdam proved that the "king" bee in fact is a queen.

Swammerdam was born in Amsterdam and qualified as a physician at the University of Leiden, although he never practiced medicine. In addition to his studies of insects, he made a number of discoveries in vertebrate anatomy, including that the brain and nervous system are responsible for muscle contractions. Swammerdam summarized his findings first in a book called *Historia Insectorum Generalis* (General description of insects), published in 1669, but later expanded this into *Bybel der Natuure* (Bible of nature), which included a long treatise on bees and 12 pages on ants. Swammerdam died from a fever on February 17, 1680, leaving his manuscript to a friend, but there were many delays, and the work finally appeared as two volumes in 1737 and 1738, in Dutch with a Latin translation and 53 of Swammerdam's illustrations.

FRANCESCO REDI AND SPONTANEOUS GENERATION

Jan Swammerdam rejected the notion of spontaneous generation, as did Francesco Redi (1626–97), but Swammerdam's rejection was total, allowing no exceptions, whereas Redi believed that certain animals, such as intestinal worms, could appear in this way. Spontaneous generation is the belief that under certain conditions living organisms can appear from nonliving matter. Everyone knew that after a fairly short time food becomes unfit to eat and not long after that maggots appear in it. No one placed the maggots there, so it seemed reasonable to suppose that they simply appeared by themselves. Many thatched barns had leaky roofs and when the grain stored in them became moist, before long it also became moldy. Farmers knew that when this happened they would find large numbers of mice living in nests inside the grain. It seemed obvious that the mice, which no one had placed there, had been produced spontaneously from the moldy grain. The Flemish physician Jan Baptista van Helmont (1577–1644) even had a recipe for producing mice in this way. He said that if sweaty underwear were to be placed in an open jar together with a quantity of wheat grain, moisture from the underwear would penetrate the grain, and the grain would turn into mice after about three weeks. Male and female mice would be present, he said, and they would reproduce normally.

Francesco Redi set out to discover once and for all whether maggots and flies can arise spontaneously in putrefying meat. In doing so he performed what may have been the world's first properly controlled scientific experiment. Redi believed that flies do not appear spontaneously, but only from eggs laid by a previous generation of flies.

To test his hypothesis he took a number of jars and placed a piece of fresh raw meat into each jar. One group of the jars he left open. A second group of jars he sealed to make them fully airtight. A third group of jars he sealed by covering them with gauze, fastened tight around the rim. There were several jars in each of his three groups. He then watched to see what would happen. He observed flies entering the open jars and walking on the meat; some time later maggots appeared in these jars. No maggots appeared in any of the tightly sealed jars. Flies could not pass through the gauze sealing the third group of jars, but they did settle on top of it, laying eggs that hatched into maggots on top of the gauze. Redi had demonstrated that maggots cannot emerge spontaneously from meat or any other material, and nor do they emerge from anything carried invisibly through the air because air was able to enter the gauze-covered jars. They occur only where adult flies have direct access to the material.

Convincing though Redi's experiment was, belief in the spontaneous generation of microorganisms persisted until well into the 18th century. Until the formulation of the theory of evolution (see "Charles Robert Darwin and Evolution by Means of Natural Selection" on pages 139–142), there seemed to be no other way to account for the appearance of the first members of each species.

Francesco Redi was born on February 19, 1626, at Arezzo, Italy, and studied medicine and philosophy at the University of Pisà. He qualified as a physician in 1647 and was employed as a physician at the court of the grand dukes of Tuscany. He was also a poet. Redi died on March 1, 1697, at Pisà.

DESCARTES AND MECHANICAL ANIMALS

At the same time as naturalists were dissecting animals and learning more and more about the way that their bodies functioned, other scholars were exploring the physical and chemical world. They believed they were revealing the innermost workings of the atmosphere, oceans, and the Earth itself, and the picture they saw emerging was of a huge machine. Robert Boyle (1627–91), a founding member of the Royal Society and one of Europe's leading scientists, proposed that all physical objects consist of minute moving particles. When those particles encounter a human body the person perceives their effect as a sensation of sight, heat, sound, taste, or smell. The sensation is produced in the brain, the particles possessing only such properties as size, shape, and motion. With his discovery of the action of gravity and the laws of motion, Sir Isaac Newton (1642–1727) portrayed the solar system as a machine.

This mechanistic view of the world also affected the way many naturalists regarded animals, and no one expressed it better or more forcefully than René Descartes (1596–1650), one of history's greatest philosophers. In 1637 Descartes published a book with the title *Discourse on the Method of Rightly Conducting the Reason, and Searching for Truth in the Sciences.* This was an outline of a philosophical system he had been developing privately, and it was based on mathematics. Descartes was a skilled mathematician and recognized that since mathematical reasoning is the only known route to certainty all reasoning should employ mathematical methods. He based this on four principles:

- The philosopher should accept as true only those assertions that cannot logically be denied—for example, "a father is older than his son."
- Every question should be divided into manageable parts.
- In seeking answers, the philosopher should commence with the simplest matters and advance to the more complex ones.
- Finally, the philosopher should pause to review the progress made sufficiently often to be able to retain a clear impression of the overall argument.

His mathematical reasoning led Descartes to conclude that every aspect of the physiology and behavior of nonhuman animals could be explained without supposing that they are thinking or are conscious. They are, in fact, machines, albeit exceedingly complex machines. The following excerpts from the *Discourse* explain this.

> . . . this will not seem strange to those, who, knowing how many different automata or moving machines can be made by the industry of man, without employing in so doing more than a very few pans in comparison with the great multitude of bones, muscles, nerves, arteries, veins, or other pans that are found in the body of each animal. From this aspect the body is regarded as a machine which, having been made by the hands of God, is incomparably better arranged, and possessing in itself movements which are much more admirable, than any of those which can be invented by man.
> . . . if there were machines which bore a resemblance to our body and imitated our actions as far as it was morally possible to do so, we

should always have two very certain tests by which to recognize that, for all that, they were not real men. The first is, that they could never use speech or other signs as we do when placing our thoughts on record for the benefit of others. . . . And the second difference is, that although machines can perform certain things as well as or perhaps better than any of us can do, they infallibly fall short in others, by the which means we may discover that they did not act from knowledge, but only from the disposition of their organs.

Descartes went on to argue that although parrots and magpies can be taught to speak, they cannot hold a conversation the way any person can, and although monkeys may be highly dexterous in performing certain tasks, they lack the versatility of any human. This demonstrated that only humans possess reason. Interestingly, this is also very similar to the Turing test, devised in 1950 by the English mathematician Alan Turing (1912–54) to distinguish between a live person and a computer.

It was a bold vision, because Descartes and his contemporaries lived at a time when machines were powered only by clockwork, gravity, wind, or springs and levers. It was also a powerful vision that succeeded in stripping away the mysticism and the last vestiges of mythology surrounding the study of natural phenomena, including animals. Animals were very complex machines, and the naturalists were nowhere near being able to explain them, but in principle it was possible to understand them.

René Descartes was born on March 31, 1596, at La Haye en Touraine in central France. In 1802 the town was renamed La Haye-Descartes, and in 1967 the name was changed again, to Descartes. His father, Joachim, was a councillor in the parliament of Rennes, in Brittany. When he was eight René was sent to the Collège Royal Henry-le-Grand, a Jesuit school at La Flèche, on the banks of the River Loire, where he spent five years studying literature and grammar followed by three years studying science, philosophy, and theology. Mathematics was his favorite and best subject. He entered the University of Poitiers in 1612 to study law, graduating in 1616. He wished to see the world, so in 1618 he went to the Netherlands to volunteer for the army of Prince Maurice of Nassau and became a military engineer. He left the army in 1619 and continued his travels through Europe, finally settling in the Netherlands in late 1628. He remained there until 1649, when he accepted an invitation from Queen Christina of Sweden to teach her mathematics. Descartes had never been physically strong and seldom rose before noon. Even as a schoolboy he had been permitted to remain in bed until 11, and in later life he did most of his work in bed. On his arrival in Stockholm, however, he learned that the queen wished to receive her instruction at five o'clock every morning. Rising so early and walking to the palace in the intense cold of the Swedish winter proved too much for him. He fell ill and died from pneumonia on February 11, 1650.

LINNAEUS AND THE BEGINNING OF MODERN CLASSIFICATION

By the beginning of the 18th century, naturalists had access to lists of animals compiled from specimens and reports from all over the world. Specimens from abroad usually arrived preserved in alcohol or as skins and skeletons. After examination the animals were often stuffed for display. There was still no universally accepted system for naming animals or plants, however, and extracting information from the ever-growing lists was becoming increasingly difficult. The solution to the problem appeared in 1735, in the form of a book of only 11 pages. The book's full title was *Systema naturae per regna tria naturae, secundum classes, ordines, genera, species, cum characteribus, differentiis, synonymis, locis* (System of nature through the three kingdoms of nature, according to classes, orders, genera, species, with characters, differences, synonyms, places). The book was written in Latin and published in the Netherlands. Its author was a Swedish botanist Carl Linnaeus (see the sidebar "Carl von Linné: Binomial Nomenclature" on page 124). *Systema naturae* was an immediate success. Further editions appeared in subsequent years, in each of which Linnaeus added more names. By the 13th and final edition, published in 1770, the original 11 pages had grown to about 3,000.

Linnaeus, in common with all the scholars of his day, believed that all species had been created by God and that their essential characteristics were fixed, although there were many minor variations within each species. Every species corresponded to an ideal form, rather like Plato's Theory of Forms (see "Plato and the World of Ideas" on pages 15–18), and could not change. This fixity made it possible to arrange species in hierarchies based on their structure

CARL VON LINNÉ: BINOMIAL NOMENCLATURE

Carl Linnaeus was born on May 23, 1707, on a farm at Råshult, in Småland, southern Sweden. His father, Nils Ingemarsson Linnaeus, was a clergyman who, as was the custom at the time in Småland, had originally been known by his patronomic name, Ingemarsson, meaning son of Ingemar. But when Nils enrolled at university the authorities needed to know his surname, so he gave himself the name Linnaeus, after an old and much-loved lime tree (*linn* in the Småland dialect) on land the family owned. Nils was a keen gardener and familiar with the wild flowers growing in the area, and by the time Carl was four his father was teaching him about plants and encouraging him to learn their names. His resulting passion for naming plants came to shape Carl's adult life.

Carl began his education in 1714, when his parents appointed John Telander, the 20-year-old son of a local farmer, as a tutor. This did not go well and after two years Carl was sent to school in the nearby town of Växjö. Conditions there were little better. The teachers were brutal, and Carl was a poor student, although he did learn Latin and, as he grew older and

had more freedom, developed his interest in plants. In 1727 Linnaeus enrolled at the University of Lund as a medical student, transferring the following year to the University of Uppsala. Uppsala had a botanical garden, where Linnaeus spent as much time as he could. That is where he met Olof Celsius (1701–44), the dean of Uppsala Cathedral, who introduced him to the garden's director, Olof Rudbeck (1660–1740). Rudbeck was near retirement and seeking someone to take over from him. He invited Linnaeus, then still a first-year student, to deliver botany lectures. At about this time Linnaeus began to suspect it might be possible to devise a system for classifying and naming plants on the basis of their flower structure.

In 1732 Linnaeus went on an expedition to Lapland, then a vast wilderness, almost unknown to most Swedes, extending across northern Norway, Sweden, Finland, and Russia. Linnaeus spent four months there, from May to September, walking all the way to the shores of the Arctic Ocean collecting plants, among which he found about 100 species that were new to science.

and appearance. Linnaeus constructed a hierarchy based on three kingdoms, which he called *Regnum animale* (animal kingdom), *Regnum vegetabile* (plant kingdom), and *Regnum lapideum* (mineral kingdom). He included minerals because naturalists believed these had distinct forms that could be classified in the same way as plants and animals.

The *Systema naturae* grouped animals into six classes: Quadrupedia (mammals), Aves (birds), Amphibia (amphibians and reptiles), Insecta (insects), and Vermes (all invertebrates other than insects). In addition, Linnaeus included a category he called Paradoxa, of animals

Linnaeus was also interested in mineralogy. He lectured on this subject in 1733, and the following year he visited the county of Dalarna to visit a copper mine. While there he met a local doctor Johan Moraeus and his daughter Sara Elisabeth. Linnaeus and Sara became engaged and were married in 1739. Moraeus had gained his medical qualification in the Netherlands and he persuaded Linnaeus to complete his medical studies there. Linnaeus qualified as a physician in 1735 at the University of Harderwijk, then moved to Leiden. While in Leiden, Linnaeus showed a manuscript he had written to the botanist Jan Fredrik Gronovius (1690–1762). Gronovius was so impressed he published it at his own expense, with the title *Systema Naturae*. In 1737 Linnaeus published *Genera Plantarum*.

At the age of 31, Linnaeus set himself up to practice medicine in Stockholm. He was a successful physician and also admired for his scientific work. He became a founding member and first president of the Royal Swedish Academy of Sciences in 1739, and in 1741 he was appointed professor of practical medicine at the University of Uppsala. Within a year he had exchanged this post for that of professor of botany. He published *Species Plantarum* in 1753. It listed all the species of plants known at the time and classified them according to the system he had devised. This and *Genera Plantarum* are considered his most important works and to this day they remain the starting point for the naming of all flowering plants and ferns, based on his system of *binomial nomenclature*. In total, Linnaeus wrote about 180 books.

In 1751 Linnaeus had begun writing accounts of the royal collection of natural history objects— mostly zoological—and eventually he compiled full catalogs of them. This brought him into frequent contact with the king and queen. In 1757 King Adolf Fredrik ennobled him and the privy council confirmed this in 1761. Linnaeus then became Carl von Linné.

Linnaeus's health began to fail in 1772, and in 1774 he suffered a stroke. After making a partial recovery, he suffered a second stroke in 1776, which paralyzed his right side. He died on January 22, 1778, during a ceremony in Uppsala Cathedral, where he is buried.

that defied his classification, including hydra, satyr, and phoenix. He divided the classes into orders and the orders into genera and species. In the class Quadrupedia, defined as those animals having body hair and four feet, producing live young and suckling them with milk, he placed the orders Anthropomorpha (monkeys, apes, humans, and sloths), Ferae (carnivores), Glires (hyraxes, squirrels, hares, shrews, and rodents), and Pecora (camels, deer, goats, sheep, and cattle).

In the 10th edition of his work, published in 1758, Linnaeus included the whales and dolphins in the Quadrupedia, and it was also in this edition that he introduced his method of using two names—

MODERN CLASSIFICATION

Zoologists classify animals according to rules set out in the International Code of Zoological Nomenclature. Modern biological classification begins by dividing organisms into three domains: Bacteria, Archaea, and Eukarya. The domain Eukarya contains all those organisms with cells that are enclosed within membranes and that contain membrane-bound bodies with specific functions. The Eukarya includes the kingdoms Animalia (also called Metazoa), Plantae (also called Metaphyta), Fungi, and Protista.

All animals belong to the kingdom Animalia. The kingdom is subdivided, in descending order, into: subkingdom, phylum, subphylum, class, subclass, superorder, order, suborder, infraorder, family, subfamily, tribe, genus, subgenus, species, subspecies. Not all of the categories may be needed to fully describe a particular species. For example, the wolf is classified as:

Class: Mammalia
Subclass: Theria
Infraclass: Eutheria
Order: Carnivora
Family: Canidae
Genus: *Canis*
Species: *lupus*

binomial nomenclature—for animals (he had applied it to plants in the 1753 edition of his book *Species plantarum*). The binomial system was quickly accepted, and today all biologists use it (see the sidebar "Modern Classification" above).

As the name indicates, every organism has at least two names, conventionally printed in italic type. The first name, conventionally printed with a capital initial, is that of the genus. The second name—known as the specific or trivial name—is that of the species.

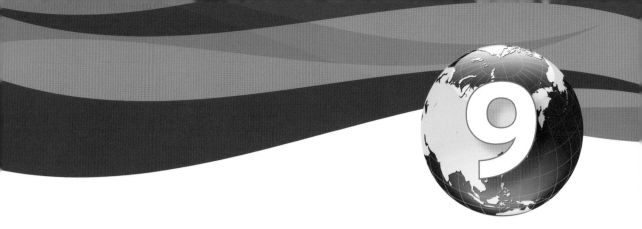

Generation, Species, and Evolution

It had been obvious from the earliest studies of animals that each species was adapted to the circumstances in which it lived. At first, it seemed that every animal had been created for a particular way of life under particular environmental conditions. The overall picture was very static, but in the course of the 18th century naturalists found this idea increasingly difficult to defend. When animals that appeared to be fitted for life in one part of the world were transported to another, with a very different climate, they often flourished in their new surroundings. Animals were much more adaptable than the static picture suggested.

The recognition that most animals could adapt to new conditions opened up a whole new field of study as natural philosophers sought to discover how animals adapted. This led them to explore the ways in which characteristics that equipped an animal for its way of life were passed from one generation to the next, and that line of inquiry led to studies of reproduction and, eventually, to theories about the origin and evolution of species. This chapter explores some of the steps along the road to the modern theory of evolution by means of natural selection.

PREFORMATION

In spring, male frogs seek out females, waiting for them in the still, shallow water of small ponds and ditches. The females lay eggs in the water, which the males fertilize, and the fertilized eggs remain as frogspawn, a mass of transparent eggs, each with a black nucleus. In time the nucleus grows a tail, and eventually the tadpole emerges from the egg, swimming freely. In time it grows legs, its tail disappears, it becomes a froglet—a tiny frog.

This entire process is visible and familiar, but it raises a question. Does the froglet develop because of radical changes in the material from which the egg nucleus is made, or was the froglet there all the time, but hidden? In the case of the frog the answer seems obvious, but is that the case for all animals? The theory that animal embryos develop through fundamental changes in the substances from which they are made is called *epigenesis*. The theory that they develop through the unfolding of characteristics that were there from the time of fertilization is called *preformation*.

In the late 18th and early 19th centuries, the concept of evolution referred mainly to preformation. It described the way the features of the developing, or evolving, animal revealed themselves. This can cause confusion, because today evolution means something much more like epigenesis, but applied to entire species rather than individual animals.

For a very long time no one knew which theory was correct, but when Antoni van Leeuwenhoek (see "The Invention of the Microscope" on pages 112–116) observed and drew spermatozoa, those favoring preformation, the preformationists, felt vindicated. The spermatozoa, and van Leeuwenhoek observed them from about 30 animal species, were clearly physical entities, and alive. They moved about. It was, therefore, impossible that embryos could develop through some kind of chemical rearrangement, as the epigeneticists supposed. It is what van Leeuwenhoek believed. He had no doubt that the spermatozoa possessed blood veins and nerves, that they were, in effect, minute versions of the adults they would eventually become. Nicolaas Hartsoeker (1656–1725), one of van Leeuwenhoek's students and his assistant, went further. In 1694 he suggested that inside each spermatozoon there was a tiny animal curled up. He called them "little infants" if they were human and "little animals" if they were not, and he drew pictures of them. These were later called *homunculi* and *animalcules*, respectively.

Hartsoeker did not claim to have seen these beings, but proposed them out of necessity—his preformationist understanding of reproduction required their existence, and with the rapid advances being made in optics he had little doubt that before long they would become visible to microscopists. The French philosopher Nicolas Malebranche (1638–1715) suggested it was not unreasonable to suppose that each homunculus and animalcule contained within itself an untold number of more homunculi or animalcules. If that were the case, every individual person or animal would be the latest version to appear of an unbroken sequence of identical beings extending back to the creation of the species and would contain within more versions yet to be born. When the stock of homunculi or animalcules was exhausted, the line would die out.

The debate between epigeneticists and preformationists continued well into the 19th century, and a version of it continues still. The German biologist August Friedrich Leopold Weismann (1834–1914) proposed that heritable characteristics flow in only one direction, from the genetic material inherited from parents to the body of the offspring. In 1892, in his book *Das Keimplasm* (Germ plasm), Weismann described how his own research had driven him to abandon his epigenetic view and accept the truth of preformation. Others disagreed. Today, there are still those who maintain that every aspect of a person, including his or her attitudes and behavior, is genetically predetermined. That is a version of preformation. Others hold that people are shaped by their environment and the way they are raised. That is a version of epigenesis.

CHARLES BONNET AND HOW APHIDS REPRODUCE

Some species of aphids, also known as plant lice, are serious plant pests. They are small, soft-bodied insects that have a strong *rostrum*—a snoutlike projection between the first pair of legs—with which they pierce plant tissues and suck up the sap on which they feed. A few aphids on a plant present no problem, but once they have found a suitable food source, the aphids multiply until they completely cover much of the plant. As well as robbing the plant of nutrients, they can transmit viruses that cause diseases. The illustration on page 130 shows a greatly magnified picture of an aphid *nymph*—the immature form—of a rose aphid (*Macrosiphon rosae*) that is about to pierce the rose leaf on which it is standing.

Aphids are superbly adapted to their way of life, and they are of practical interest to biologists because many species, though not all,

A scanning electron micrograph of the nymph (immature form) of a rose aphid (*Macrosiphon rosae*) feeding on a rose leaf. Aphids, also called plant lice, pierce plants to feed on the sap they suck from them. This aphid is about to pierce the leaf with its rostrum. *(Andrew Syred/Science Photo Library)*

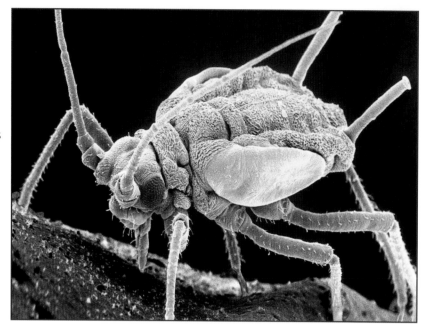

are capable of devastating plant crops. They are able to do so because of their ability to multiply rapidly, and the person who discovered their secret was the Swiss naturalist Charles Bonnet (1720–93). It provided him, or so he thought, with clear evidence for preformation.

Bonnet, a keen *entomologist* (a biologist who specializes in the study of insects), performed a number of experiments that revealed the reproductive cycle of aphids. In the fall, females and males mate and the females lay eggs that lie dormant in the soil throughout the winter. The following spring, the eggs hatch, producing nymphs that immediately begin looking for a food plant. Since the previous generation inhabited food plants, the nymph is unlikely to have to look very far. All of these nymphs are female and wingless; they are called *fundatrices* (sing. *fundatrix*). When a fundatrix finds a food plant and begins to feed, the unfertilized eggs in its body begin dividing, producing a second generation of nymphs that are born alive, but without the parent having mated. Each of these second-generation nymphs also produces young by *parthenogenesis*—reproduction without mating—and the second generation is followed by several more generations, each generation consisting only of females. Sometimes aphids produce a generation of females with wings that migrate to another plant nearby, where they start a new colony. The final generation of the season, produced parthenogenetically like its predecessors, comprises males as well as females, and both genders possess wings. They mate, the females lay eggs, and then the insects die.

An individual aphid lives for three to six weeks and is able to produce live young as soon it emerges. Consequently, between spring and fall a single fundatrix can give rise to many generations of females, and millions of descendants. Bonnet observed this rapid, parthenogenetic reproduction and assumed that each aphid contained within its body all the aphids to which it gave birth. This looked like clear evidence of preformation. Of course, it was not. The young developed from eggs, but no fertilization was involved and the development took place inside the female's body.

Charles Bonnet was born in Geneva on March 13, 1720. He trained as a lawyer but entomology was his passion. He submitted his study of aphid reproduction to the Swiss Academy of Sciences in 1740 and was admitted as a corresponding member. The following year he began to study reproduction and the regeneration of lost body parts in other animals, and in 1742 he began to study respiration in butterflies. He was elected a fellow of the Royal Society in 1743 and in the same year received the degree of doctor of laws. Bonnet published his entomological researches in 1745, in his book *Traité d'insectologie* (Entomological treatise), in which he also explained his views on the hierarchical arrangement of beings. Bonnet then turned his attention to plants, but by the 1750s his eyesight was failing, forcing him to abandon biological research. Instead, he concentrated on philosophy and psychology. He never left Switzerland and spent his last 25 years living quietly at his home at Genthod, near Geneva, where he died on May 20, 1793.

ABRAHAM TREMBLEY AND THE POLYP

Abraham Trembley (1710–84), a cousin of Charles Bonnet, was also a Swiss biologist, and in 1741 he made a discovery that challenged the concept of preformation. Trembley studied freshwater hydra, known in his day as polyps, starting with the common species *Chlorophyta viridissima*. These are small, cylindrical animals, up to about 0.4 inch (1 cm) long and rarely more than 0.04 inch (1 mm) wide, that live in most rivers and ponds where the water is unpolluted. The diagram on page 132 shows the structure of a hydra. The main part of the body, called the body stalk, contains a *gastrovascular cavity*—a space with a complex network of channels that circulate nutrients and blood. At one end there is a basal disc by which the hydra

A hydra attaches itself to a solid surface by its basal disc. The main part of its body, the body stalk, contains the gastrovascular cavity, made up of channels for the circulation of nutrients and blood. Its dangling tentacles trap food particles and direct them to the mouth. This hydra is reproducing by means of an asexual bud.

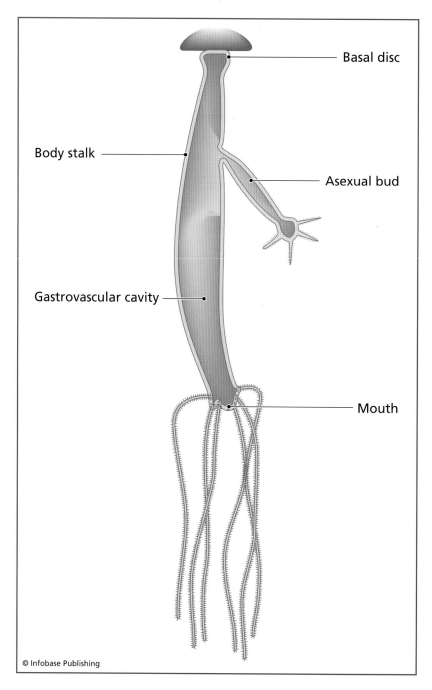

Basal disc

Body stalk

Asexual bud

Gastrovascular cavity

Mouth

© Infobase Publishing

attaches itself to a solid surface. At the other end there is a ring of tentacles, usually six, that trap smaller animals and direct them to the mouth at the center of the ring. Hydra reproduce by budding. An asexual bud appears as a protrusion on the side of the body stalk and grows into a small replica of the adult hydra before breaking free and commencing its independent life. In this drawing the young hydra is about to break free.

Trembley was not the first naturalist to study hydra. Antoni van Leeuwenhoek (see "The Invention of the Microscope" on pages 112–116) had also described them, but thought they were plants. Trembley recognized that they are animals, and he observed their power of regeneration. He found that if he sliced a hydra into two parts, each part grew its missing half and became a complete hydra. Bonnet studied the ability of some animals to regenerate body parts that have been lost, such as claws in some crustaceans and tails in some lizards, but Trembley observed the replacement of half of an animal's body. He studied three different species of hydra and published his findings in 1744 in a book entitled *Mémoires, pour servir à l'histoire d'un genre de polypes d'eau douce, à bras en forme de cornes* (Memorandum, to serve as the history of a genus of freshwater polyps, with arms in the shape of horns). A German translation appeared in 1791.

Trembley's discovery had profound implications, because the missing part of the body was very obviously growing out of the surviving part. This was clearly an example of epigenesis, and many naturalists seized on it to assert their belief that matter itself possessed the property of generating living beings. If they were right, it meant that far from being static, with everything—and everyone—occupying a preordained position in a hierarchy, the universe was unpredictable and constantly changing. It was more than a biological challenge; it was, quite literally, a revolutionary challenge to any idea of a static social order.

Abraham Trembley was born in Geneva on September 3, 1710, the son of a politician and soldier, who, in 1733, found himself on the wrong side during an uprising. He lost his position and the family went into exile. Abraham had studied mathematics, but now was forced to seek employment and he moved to the Netherlands where he became a tutor to the children of wealthy families. In 1736, while teaching the children of Count William Bentinck at Sorgvliet, the Bentinck mansion near The Hague, Trembley discovered hydra (*Chlorophyta viridissima*) in a stream flowing through the estate.

Trembley left Sorgvliet in 1747 and spent some years traveling around Europe. The Royal Society elected him a fellow and awarded him its Copley Medal in 1743. Trembley died in Geneva on May 12, 1784.

COMTE DE BUFFON AND HIS NATURAL HISTORY

The 18th century was a time of revolution, in ideas as well as in systems of government, and one of the most popular books about animals expressed some very radical views. Its author maintained that if two animals are related, for instance lions and tigers, this must mean that some kind of link connects them, just as two people who are cousins are related because they share one set of grandparents. Humans and apes are similar in many ways, so they, too, must be linked, perhaps by a shared ancestor. Linnaeus (see "Linnaeus and the Beginning of Modern Classification" on pages 123–126) had categorized species as being related simply because of their physical resemblance, placing lions and tigers together because both are large catlike animals.

The book with the radical ideas bore the title *Histoire naturelle, générale et particulière* (Natural history, general and particular). The first of its 36 illustrated volumes appeared in 1749, the last in 1804. Eight further volumes were added after the author's death. The work was translated into many European languages and became very famous.

The author was Georges-Louis Leclerc, who was ennobled in 1773, becoming the comte de Buffon. Buffon was born on September 7, 1707, at Montbard, in the south of France. His mother inherited a large fortune and the Buffon estate, and in 1717 the family moved to Dijon. Buffon was educated at a Jesuit college in Dijon and then at the University of Angers, where he studied mathematics and science but had to leave in 1730 after he became involved in a duel. He went to live in Nantes, then toured Europe, returning to France in 1732 on hearing of his mother's death. He inherited his mother's fortune and could afford to devote himself wholeheartedly to mathematics, botany, and forestry. In July 1739 Buffon was appointed keeper of the Jardin du Roi, the royal botanical garden in Paris, which he transformed into a major research institution. He held this position until his death on April 16, 1788. Buffon was made a fellow of the Royal Society in 1739 and a member of the French Academy of Sciences in 1753.

Buffon intended his *Histoire Naturelle* to be an encyclopedia describing everything that was known about the natural world. He enlisted the help of many eminent natural historians and edited their contributions into a stimulating and very readable work. He proposed that relationships among species represented either their improvement or degeneration from their shared ancestor, and he revived the idea of spontaneous generation to explain the origin of the ancestral species.

JEAN-BAPTISTE LAMARCK AND ADAPTIVE EVOLUTION

In 1778 Buffon read and approved of a three-volume work entitled *Flore Française* (French flora—*flora* means all of the plants found in a specified area; in the case of a book it is a list of those plants). The flora had been compiled by Jean-Baptiste Lamarck (1744–1829), an impoverished aristocrat and former soldier who had become interested in meteorology and botany. *Flore Française* won him membership of the Academy of Sciences in 1779, and in 1781 Lamarck was appointed botanist to King Louis XVI. With Buffon's support he spent the years from 1781 to 1785 traveling around Europe and on his return he took up an appointment as assistant botanist at the Jardin du Roi, working under Buffon. During the French Revolution, the "King's Garden" became an unfashionable name, and in 1791 Lamarck changed it to Jardin des Plantes. Lamarck was also a zoologist, and in 1793, when the Jardin des Plantes was once more reorganized and became the Muséum national d'Histoire naturelle (National Museum of Natural History), to be managed by 12 professors, he was appointed its professor of zoology. In fact, his academic field comprised the insects and worms—the groups Linnaeus classified as Insecta and Vermes.

Lamarck familiarized himself with the insects and worms, which most previous naturalists had thought unworthy of serious study, and lectured on them. He later called them the Invertebrata. Biologists no longer recognize the Invertebrata as a taxonomic group, but the word *invertebrate* that Lamarck coined remains in the language. He separated the Crustacea, Arachnida (scorpions, spiders, and mites), and Annelida (segmented worms) from the Insecta, and the sea squirts (urochordates) and barnacles from the Mollusca (barnacles are crustaceans). Lamarck was also the first person to recognize that living organisms need to be studied as a whole, and he coined the word *biology* to describe that branch of science.

In 1809, Lamarck published *Philosophie zoologique* (Zoological philosophy), the first of his books on zoology. He published a seven-volume work, *Histoire naturelle des Animaux sans vertèbres* (Natural history of animals without vertebrae), between 1815 and 1822. Lamarck pointed out that nothing can live unless its constituent parts are cells or are made from cellular tissue.

As part of his work on invertebrates, Lamarck made extensive studies of fossils. He classified them, and in doing so he observed the ways animals had changed over long periods of time. Like Buffon, he saw that species were not fixed. They could change, but Lamarck went much further than Buffon in proposing a process by which this could occur. He thought that parts of the body would develop or wither and disappear according to whether or not they were necessary to the life of the animal possessing them. Lamarck expounded the implications of this in *Philosophie zoologique* as two laws:

- **First law.** Changes in the environment require animals to change their behavior in order to satisfy their needs. Their altered behavior leads to greater or lesser use of particular parts of their bodies. Changed use causes those parts to grow larger or smaller.
- **Second law.** All such changes in the relative sizes of body parts are heritable. Consequently, animals change physically over many generations as they adapt to the environmental conditions in which they live.

This was Lamarck's theory of adaptive evolution. It is now usually known as Lamarckian evolution, or the inheritance of acquired characters. Lamarck did not believe species could become extinct; they simply changed, one species gradually evolving into its successor. He saw this as a progressive process tending to greater and greater perfection and complexity. In *Philosophie zoologique* he wrote: "Nature, in producing in succession every species of animal, and beginning with the least perfect or simplest to end her work with the most perfect, has gradually complicated their structure."

As animals evolved into ever-greater complexity, a constant supply of the simplest organisms was needed to maintain the process. Lamarck believed that these organisms appeared by spontaneous generation.

Jean-Baptiste-Pierre-Antoine de Monet, chevalier de Lamarck, was born on August 1, 1744, in Bazentin-le-Petit, a village in Picardie in northern France. He was the 11th child, and the family meant him to enter the church. In about 1756 he entered a Jesuit seminary at Amiens. Lamarck's father died in 1760, however, and the following year Lamarck joined the army. He fought with distinction in the Seven Years' War and was made an officer. The war ended in 1763, and Lamarck was posted to a garrison in the south of France, but his health deteriorated and he left the army in 1766 with a pension that was too small to support him. He worked for a time as a bank clerk while he studied medicine, but after four years he abandoned medicine and turned to botany. On August 8, 1778, Lamarck married Marie-Anne-Rosalie Delaporte. They had six children. Marie died in 1792, and the following year Lamarck married Charlotte Reverdy, who died in 1797. He married Julie Mallet in 1798; she died in 1819.

In his later years, Lamarck's eyesight began to fail. Eventually he became blind and died in great poverty in Paris on December 18, 1829. His family had to apply to the Academy of Sciences for financial assistance and were forced to auction his books and the contents of his house.

ERASMUS DARWIN AND THE PROGRESS OF LIFE

Erasmus Darwin (1731–1802) was a poet, physician, philosopher, and naturalist. He was also the paternal grandfather of Charles Darwin (see "Charles Robert Darwin and Evolution by Means of Natural Selection" on pages 139–142).

Darwin's most famous poetical work was *The Botanic Garden*, a set of two long poems, *The Economy of Vegetation* (published 1791) and *The Loves of the Plants* (published 1789). The first part, *The Economy of Vegetation*, discusses a variety of scientific and technological topics. *The Loves of the Plants* guides the reader through Linnaeus's classification of plants by describing 83 species, together with long explanatory footnotes.

His most important work was *Zoönomia; or The Laws of Organic Life,* published in four volumes between 1792 and 1796. His last long poem, *The Temple of Nature,* was published posthumously in 1803. In *The Temple of Nature* Darwin describes the progress of life from its earliest beginnings to civilized society. The poem contains Darwin's ideas on the evolution of species, forming what he called "one living filament." He described it succinctly in the following lines:

> Organic life beneath the shoreless waves
> Was born and nurs'd in ocean's pearly caves;
> First forms minute, unseen by spheric glass,
> Move on the mud, or pierce the watery mass;
> These, as successive generations bloom,
> New powers acquire and larger limbs assume;
> Whence countless groups of vegetation spring,
> And breathing realms of fin and feet and wing.

Darwin proposed that new species arose through the modification of earlier species in a way very similar to the process Lamarck would describe just a few years later. So far as anyone knows, the two men were unaware of each other's work. Darwin's theory was a little different from Lamarck's, however, because it suggested that evolutionary change is driven partly by competition and sexual selection among individuals. He wrote that: "The final course of this contest among males seems to be, that the strongest and most active animal should propagate the species which should thus be improved." This places his evolutionary theory somewhere between those of Lamarck and of Charles Darwin.

Erasmus Darwin was born on December 12, 1731, at Elston Hall in Nottinghamshire, the youngest of seven children. He was educated at Chesterton Grammar School, in Derbyshire, and at St. John's College, University of Cambridge, from where he moved to Edinburgh to study medicine at the University of Edinburgh medical school. In 1756 Darwin began to practice as a physician in Nottingham, but attracted few patients and so moved to Lichfield, in Staffordshire, in 1757. Early success with a patient who was close to death brought popularity, and Darwin continued to practice successfully in Lichfield until 1781. His reputation spread so far that King George III invited him to become his personal physician, but Darwin declined.

Lichfield was the birthplace of Samuel Johnson (1709–84) and the poet Anna Seward (1747–1809). Darwin knew both of them.

Darwin married Mary Howard (known as Polly) in 1757. The couple had four sons and one daughter, but the daughter and one of the sons died in infancy. Another son, Robert Waring Darwin (1766–1848), was Charles Darwin's father. Mary died in 1770, but Darwin continued to live in the same house with his housekeeper and Robert's governess, Mary Parker, with whom he had two daughters. He married a widow, Elizabeth Pole, in 1781, and moved into her home, Radburn (now Radbourne) Hall, near Derby. In 1782 they moved into the town of Derby. The couple had seven children, including Frances Ann Violetta Darwin (1783–1874), the mother of Sir Francis Galton (1822–1911). In 1802 Darwin moved to Breadsall Priory, to the north of Derby. He died unexpectedly soon after the move, on April 18, 1802.

CHARLES ROBERT DARWIN AND EVOLUTION BY MEANS OF NATURAL SELECTION

Modern evolutionary theory is derived from the work of Charles Darwin (1809–82). Darwin worked on his theory for many years and finally published it because of a paper he received on June 18, 1858. The paper, entitled "On the Tendency of Varieties to Depart Indefinitely From the Original Type," was written by Alfred Russel Wallace (1823–1913), a naturalist working in tropical Asia with whom Darwin had been corresponding since 1856. Darwin realized that Wallace had developed a theory very similar to his own, although there were important differences. He sent Wallace's essay to his friend Sir Charles Lyell (1797–1875), asking Lyell to return it when he had read it and saying that although Wallace had not asked him to find a publisher for it, he would write to Wallace immediately, offering to send it to an appropriate journal. Darwin had no wish to harm Wallace in any way and so it was agreed that Darwin would prepare two short extracts from an earlier manuscript and part of a letter outlining his theory to Asa Gray (1810–88), professor of natural history at Harvard University. These would be read together with Wallace's paper at a meeting of the Linnean Society in London. The meeting took place on July 1, 1858. Neither Wallace nor Darwin could be present; Wallace was still overseas and Darwin was unwell. The papers were read by Lyell and by another of Darwin's friends, Sir Joseph Dalton Hooker (1817–1911).

After the meeting, Darwin continued to work on the book in which he outlined his theory. He handed the manuscript to his publisher, John Murray, later in 1858, and on November 2, 1859, he received his first copies of what was to become one of the most famous books ever published: *On the Origin of Species by Means of Natural Selection or The Preservation of Favoured Races in the Struggle for Life.* The book is more often known as the *Origin of Species.* The theory Darwin outlined can be summarized easily:

- Every member of a species is slightly different from all other members; in other words there is variation among the individuals of a species.
- On average, parents produce more offspring than are needed to replace them; in their lifetimes two parents are capable of producing more than two offspring.
- Populations cannot possibly continue to increase indefinitely and, on average, population sizes remain stable.
- The stability of population sizes despite the capacity of populations to increase implies that in every generation there must be competition for survival.
- In that competition, those individuals possessing variations that help them to obtain resources and mates will tend to produce more offspring than other individuals lacking those variations.
- The offspring will inherit their parents' advantageous traits.
- Over many generations, the descendants of the better-adapted individuals will become more numerous.
- As gradual changes in the environment alter the conditions under which species live, different variations among individuals will prove advantageous and nature will select those variations. This natural selection will lead to the emergence of new species.

Although the theory is simple, its implications are profound, and biologists are still exploring them. There have been challenges from time to time, and Darwin's original theory has been modified, but in essence it has been found to be true, confirmed by a vast accumulation of observational and experimental evidence. No scientific theory springs suddenly from nowhere, of course, and Darwin's theory is the culmination of centuries of observations and speculations by some of the greatest minds—a fact that does nothing whatever to detract from Darwin's achievement.

Charles Robert Darwin was born on February 12, 1809, in Shrewsbury, England, the fifth of the six children of Robert Waring Darwin, F.R.S., a very successful physician and financier, and Susannah Darwin, daughter of the ceramics manufacturer Josiah Wedgwood. Charles was the grandson of Erasmus Darwin. The family was wealthy. Charles entered Shrewsbury School in 1818 as a boarder, and in 1825 he began to study medicine at the University of Edinburgh. He hated medicine, finding the lectures boring and the demonstrations disgusting. So in 1828 his father transferred him to Christ's Church College at the University of Cambridge to study theology, but he did not like theology, either. Natural history was the subject that inspired him, and he met the professor of botany at Cambridge, John Stevens Henslow (1796–1861), who encouraged him. The two became lifelong friends. Darwin graduated in 1831, doing well in his examinations. He remained in Cambridge until June, then enrolled in a geology course taught by Adam Sedgwick (1785–1873) and went on a mapping expedition with him to Wales. Darwin then returned home, where a letter from Henslow was waiting for him. It suggested he apply for the position of companion to Robert FitzRoy (1805–65), captain of HMS *Beagle,* on a surveying expedition to South America. The *Beagle* sailed from Devonport, Plymouth, on December 27, 1831, and arrived back in England on October 2, 1836.

During the voyage Darwin had collected specimens and kept up a correspondence with his scientific friends in England. When he returned he learned that some of his letters had been published privately, and he was regarded as an important scientist. On November 1, 1836, he was made a fellow of the Royal Geographical Society. The official account of the *Beagle* voyage was published in four volumes in May 1838, with Darwin's *Journal and Remarks 1832–1835* forming the third volume; this was published separately in 1845 as *Journal of Researches into the Natural History and Geology of the Countries Visited During the Voyage of HMS* Beagle *Round the World, Under the Command of Captain FitzRoy, R.N.,* more often known as *The Voyage of the* Beagle. It is still in print.

On January 29, 1839, Darwin married his cousin, Emma Wedgwood (1808–96). The delighted family gave them a dowry and investments that brought them an income on which they could live in comfort for the rest of their lives. In 1842 the couple moved to Down House, in Kent, bought for them by Robert Darwin, and that is where they spent the rest of their lives.

Following his return from the *Beagle* expedition, Darwin found himself under tremendous pressure of work. He had to write and correct his *Journal* for inclusion in the official report of the voyage. He began editing and preparing the reports on his zoological collections written by various experts as well as Darwin himself. These were published between February 1838 and October 1843 in five volumes as *Zoology of the Voyage of H.M.S.* Beagle *Under the Command of Captain FitzRoy, R.N., during the Years 1832 to 1836.* He had agreed to write a book on geology, and, in March 1838, after resisting for a time, he accepted the post of secretary to the Geological Society of London. His health began to deteriorate, and he never recovered. From that time he was prone to attacks of stomach pains and vomiting and suffered from headaches, palpitations, boils, and other symptoms. The cause of his illness was never diagnosed, but the symptoms were especially severe whenever he had to attend meetings or social events, which he found stressful.

Darwin continued with his research and writing. He published *The Descent of Man, and Selection in Relation to Sex* in 1871, *The Expression of Emotions in Man and Animals* in 1872, as well as a number of books on botany, and an autobiography that was published in 1887. He died at Down House on April 19, 1882. His funeral was held in Westminster Abbey, where Darwin is buried.

Genetics and Relationships

By the end of the 19th century most biologists had come to accept it as fact that plant and animal species evolve from earlier species. Lamarck had suggested one process by which evolution might occur, but Darwin had proposed an alternative that was more powerful and more convincing. The theory was still incomplete, however, because no one knew how characteristics pass from parents to their offspring. It is obvious that offspring resemble their parents, but why do they?

This chapter tells how the puzzle was solved. The story begins in the middle of the 19th century with the story of an Augustinian monk who performed breeding experiments with pea plants and made a major discovery that was lost for decades. The chapter then records the discovery of the *gene*—the basic unit of heredity—and of its structure, the famous double helix. Genes are units of code that carry instructions for the synthesis of *amino acids*—the basic building blocks of proteins. It follows, therefore, that genes encode for the suites of proteins that determine the physical form and function of the organism carrying them.

Once it became possible to examine the genetic composition of any species, comparing their genetic structures provided a measure of the relationships between species. The chapter ends by describing the way biologists use genetic relationships to examine, and where necessary revise, animal classification.

GREGOR MENDEL AND HIS PEAS

Magnificent though Darwin's evolutionary theory was in its power and elegant simplicity, Darwin had not the slightest idea about the mechanism that produces the variations on which his theory depended or how those variations pass from parents to their offspring. It was a puzzle, but a puzzle that was partly solved in 1865, just a few years after the publication of the *Origin of Species,* although neither Darwin nor any of his friends and associates knew of it.

This discovery was made by Gregor Mendel, who was a monk in the Augustinian monastery of St. Thomas in what was then the town of Brünn, in Moravia, a part of Austria; the town is now Brno, in the Czech Republic. Mendel was born on July 22, 1822, on a farm at Heizendorf, Silesia, then part of Austria; it is now Hynčice, Czech Republic. He was given the name Johann. His parents, Anton and Rosine, sent him to the local school, where his teachers quickly recognized his talent. In 1841 he began to study to become a teacher, but a farming accident in 1838 had left Anton unable to work and there had been several poor harvests. The financial situation was grave and Mendel had to leave school. He might have returned to help run the farm, but then he would have been another mouth to feed. Instead, in 1843 he entered the monastery, taking the name Gregor, and began preparing to enter the priesthood. He was ordained in 1847, but even while preparing he had continued to study science, for which the monastery had a reputation. Mendel taught Greek for a short time in 1849, but in 1850 he failed the examination to become a teacher.

Brünn was a center of the textile industry, but the factories had to import high-quality wool from Spain, and there was great interest in the possibility of improving the local sheep breeds. Mendel thought he might be able to help and planned a series of breeding experiments, initially using mice, but when this proposal proved unacceptable to the abbot, he made do with pea plants (*Pisum sativum*). He chose peas because their flowers are enclosed by the petals and ordinarily the plants pollinate themselves, but he was able to tease the petals apart and use a fine paintbrush to collect pollen that he transferred to the carpels of other plants from which he had removed the stamens before they produced pollen. In this way he could cross the plants, retaining rigorous control over the procedure.

At the time, most scientists believed in *blending inheritance*—the theory that offspring inherit the traits of both parents and these merge to form an intermediate condition. Mendel set out to test this using some pea plants that reliably bore purple flowers and others that reliably bore white flowers, a second set of plants that produced either smooth peas or peas with wrinkled skins, and a third group that produced long or short stems. If blending inheritance were true, the offspring from cross-breeding should have had pale purple flowers, peas with just a few wrinkles, and stems of intermediate length. In fact, using a very large number of plants, Mendel found that in the first generation all the flowers were purple and in the second generation there were 705 plants with purple flowers and 244 with white flowers. The experiment with smooth and wrinkled peas and long and short stems produced similar results.

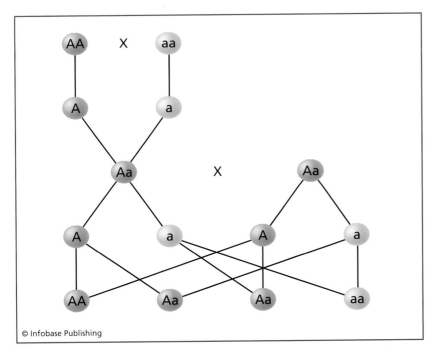

© Infobase Publishing

Two members of the same species both breed true; one is red, the other green. They breed true because they both carry identical alleles (*AA* and *aa*) for color. *A*, which is dominant, produces red; *a*, which is recessive, produces green. When the two organisms are crossbred, the first generation is all red (*Aa*). When *Aa* and *Aa* are crossed, the alleles separate, and in the second generation three are red (*AA, Aa, Aa*) and one is green (*aa*).

Gregor Mendel (1822–84) is seen holding a plant and perhaps discussing it with his fellow monks. The original of this photograph is held at the Mendelium museum, at the Abbey of St. Thomas, in Brno, Czech Republic, where Mendel was the abbot. *(James King-Holmes/ Science Photo Library)*

Translated into modern terms, Mendel had shown that blending inheritance was mistaken. Instead, every plant or animal carries hereditary instructions (now called genes) for at least two versions for each of its physical characteristics; these versions are called *alleles.* For each of its characteristics, offspring inherit one allele from each parent. In many cases the alleles are identical, but in others they are not. Where they differ, one allele is dominant (call it *A*) and the other recessive (call it *a*). The dominant allele (*A*) is the one that will be expressed in the offspring that inherits it. The recessive will be expressed only in those offspring that inherit two copies (*aa*). The diagram on page 145 shows the consequences of this type of inheritance. Suppose two individuals breed, one of them *AA* and the other *aa*. Their offspring will inherit one allele from each parent, so they will all be *Aa,* and *A* will be expressed. In the diagram, *A* is assumed to confer the color red and *a* to confer green, so in the first generation all the offspring will be red. If the *Aa* members of the first generation breed, each will supply the offspring with one *A* and one *a* allele. In the second generation, therefore, one offspring will be *AA*, one will be *aa*, and two will be *Aa*. Since *A* is expressed, 75 percent of the offspring in this example will be red and 25 percent will be green. This is very close to the ratio Mendel found with his purple and white pea flowers and his smooth and wrinkled peas.

The monks at St. Thomas's had formed a natural history society, and Mendel presented his findings to the society at its meetings on February 8 and March 8, 1865, as a paper "Versuche über Pflanzen-Hybriden" (Experiments in plant hybridization). The paper was subsequently published in the society's journal and copies were sent to every important scientific library in Europe and North America, but it was in German, the journal was obscure, and no one took the slightest notice. The photograph above shows Mendel at about this time.

In 1868 Mendel was elected abbot of the monastery. He remained interested in science, but the running of the monastery demanded his full attention and most of his time. He had to abandon his research. Mendel died at the St. Thomas monastery on January 6, 1884.

In 1900, three botanists were working independently on heredity, Hugo de Vries (1848–1935) in the Netherlands, Karl Erich Correns (1864–1933) in Germany, and Erich Tschermak von Seysenegg (1872–1962) in Austria. As scientists do, they began their researches by studying the scientific literature to find out what relevant work had already been done. All three of them discovered Mendel's paper, and although they were not the first scientists to read it, they were the first to recognize its importance. They, too, had found that plants inherit traits in a ratio of 3:1. The English biologist William Bateson (see "William Bateson and *Balanoglossus*" on pages 150–151) translated the paper into English; his research was with animals, using the shape of the comb in chickens, and he found that they, too, inherited traits in the same ratio.

AUGUST WEISMANN AND THE REJECTION OF LAMARCK

In 1892, the German zoologist August Weismann declared that his research had led him to reject the theory of epigenesis in favor of preformation (see "Preformation" on pages 127–129). His research also led him to refute Lamarck's theory of adaptive evolution (see "Jean-Baptiste Lamarck and Adaptive Evolution" on pages 135–137), as well as the theory of heredity Charles Darwin had proposed in his book *Variation of Plants and Animals Under Domestication,* published in 1868. The research that led Weismann to this conclusion also established him as one of the founders of *genetics*—the scientific study of inheritance.

August Friedrich Leopold Weismann was born on January 17, 1834, in Frankfurt-am-Main, Germany, where his father taught classics at a high school (gymnasium). Weismann entered the University of Göttingen in 1852 to study medicine, graduating in 1856. He held several positions as a physician and in 1863 joined the faculty of the Albert Ludwig University of Freiburg-im-Breisgau, where he taught zoology and comparative anatomy. When the university built a zoological museum and research institute, Weismann became its first director. His research began with metamorphosis in insects and reproduction in Hydrozoa, but after a few years his eyesight began to deteriorate and by the 1880s he had to abandon laboratory research because he could no longer use a microscope. Instead, he began to concentrate on biological theory and, in particular, on Darwin's theory of evolution.

Darwin had suggested that every cell in the body produced tiny particles, which he called *gemmules*. These traveled to the eggs or sperm cells carrying with them information that became incorporated in the reproductive cells. Weismann was a great admirer of Darwin, but his study of hydrozoan reproduction had persuaded him that all sexually reproducing organisms possess a substance called *germ plasm* containing the hereditary material that passes to the eggs or sperm cells. Offspring inherit the hereditary material from their parents' germ plasm, and this directs the development of the cells of the body. The germ plasm passes from one generation to the next in an unbroken line of descent, and it flows in only one direction, from the eggs and sperm cells to the body cells, and never from the body cells to the eggs and sperm. The theory Darwin had proposed—which he called *pangenesis*—would have allowed the inheritance of traits acquired by a parent during its lifetime, as Lamarck had supposed. Because it allowed heritable information to pass only from reproductive cells to body cells and never in the opposite direction, Weismann's discovery of germ plasm was incompatible with the reproduction theories of both Lamarck and Darwin.

If both parents contribute germ plasm to their offspring, however, Weismann realized that the amount of hereditary material must double with each generation. This was clearly impossible, so he suggested that cells involved in reproduction must undergo a type of cell division in which each daughter cell receives only half of the original amount of germ plasm. This type of cell division is now known as *meiosis,* and other biologists later confirmed it.

Several biologists studying cell nuclei in the 1840s had observed bodies that became more clearly visible immediately prior to cell division and that separated into two groups at division, with each daughter cell acquiring one group. Confirmation of his idea led Weismann to propose that the germ plasm is located in these bodies in the nuclei of eggs and sperm cells. The bodies absorbed the dyes used to stain material for microscopic examination, and in 1888 the German anatomist Heinrich Wilhelm Gottfried von Waldeyer-Hartz (1836–1921) gave them the name they still have: *chromosomes.*

Weismann's germ plasm theory was basically correct, but for one point. Weismann believed that variation among individuals arises through the mingling of chromosomes from both parents. This does account for some variation, but biologists now know that environmental events can affect the body in ways that alter the chromosomes, or germ plasm. This also produces variations.

In later years Weismann traveled widely, lecturing on heredity. He died in Freiburg on November 4, 1914.

WILHELM JOHANNSEN AND THE GENE

Despite Weismann's work, pangenesis did not die immediately, and Darwin's gemmules were renamed *pangenes*. Darwinian natural selection acts on variations within a species, and supporters of pangenesis believed that those variations were distributed normally. That is to say, approximately as many individuals deviate to one side of the mean as deviate to the other side. When plotted as a graph, a normal distribution produces a bell-shaped curve, as shown in the diagram below, and this kind of variation occurred in many species. Some environmental change, the pangeneticists argued, led natural selection to favor individuals on one side of the norm or the other, and after many generations this would reposition the norm, producing a new distribution curve.

This seemed entirely reasonable, but it was wrong, and the scientist who finally laid pangenesis to rest was the Danish geneticist Wilhelm Johannsen (1857–1927). Johannsen experimented with the common or French bean (*Phaseolus vulgaris*). He used pure lines—strains of an organism in which particular features appear in every generation—and he measured the size of the beans. He found that even in plants with no genetic variation, the size of the beans followed a normal distribution, which depended on the amount of food the parent plants stored in their seeds, the positions of beans in their pods, and the location of pods on the plants. It meant that the variation was not inherited and, therefore,

In a normal or Gaussian distribution (named after the German mathematician Karl Friedrich Gauss, 1777–1855), values are spread evenly about the mean.

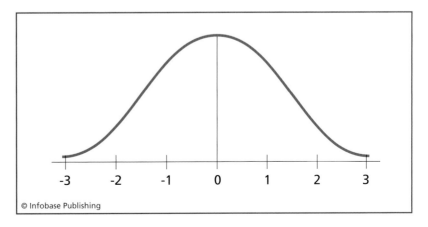

© Infobase Publishing

natural selection could not act on it. No matter how many generations of these beans he grew, Johannsen would never find the average size of the beans increasing or decreasing. He coined two terms to help him describe what was happening, *genotype*—the genetic constitution of an organism—and *phenotype*—the physical appearance of an organism. Johannsen also coined the term gene. During the 1930s, biologists drew together their understanding of natural selection and genetics to produce what they called the modern synthesis of evolutionary ideas, and it was not until then that they realized that variations must be heritable if natural selection is to act on them.

Wilhelm Johannsen was born in Copenhagen on February 3, 1857, and trained as a pharmacist in Denmark and Germany from 1872 until 1879, when he qualified. In 1881 he went to work as an assistant in the chemistry department of the Carlsberg Laboratory, where he began studying reproduction in plants. He was made a lecturer at the Royal Veterinary and Agricultural University in 1892, and later professor of botany and plant physiology. He died in Copenhagen on November 11, 1927.

WILLIAM BATESON AND *BALANOGLOSSUS*

If species evolve from earlier species, then it is likely that modern species are very different from their ancestors that lived in the distant past. Between those remote and very different ancestors and the modern species there must have been intermediate species—so-called missing links—possessing characteristics of both their ancestors and their descendants. For example, the German paleontologist Christian Erich Hermann von Meyer (1801–69) was the first person, in 1861, to describe *Archaeopteryx lithographica,* found as a fossil in Bavaria, Germany. *Archaeopteryx lithographica* is the earliest known bird, but it is unlike any modern bird in possessing jaws with small teeth rather than a bill, claws on its wings, and a long, bony tail. It lived during the Jurassic period, 199.6–145.5 million years ago, and it shares many features with the Coelosauria, a group of bipedal, carnivorous dinosaurs. It is a missing link between dinosaurs and birds—and biologists now recognize that birds are dinosaurs and not merely descended from them.

In 1883 and 1884 the English biologist William Bateson identified *Balanoglossus* as another missing link. *Balanoglossus* is an acorn worm, a marine, burrowing, wormlike animal, and when Bateson began studying acorn worms they were classed as echinoderms (phylum Echinodermata), which are invertebrate animals with spiny exteriors. Starfishes and sea urchins are echinoderms. Bateson discovered that *Balanoglossus* was different. It possessed features reminiscent of the neural chord and gill clefts of the Chordata—the phylum that contains all the vertebrate animals. It was a link between the invertebrates and vertebrates. Acorn worms are no longer classed as echinoderms, but as the phylum Enteropneusta. Bateson's discovery, made while he was a graduate student working in the United States with William Keith Brooks (1848–1908) at Johns Hopkins University, led to his election as a fellow of St. John's College, University of Cambridge.

Bateson was born on August 8, 1861, at Whitby, Yorkshire. He was educated at Rugby School and St. John's College, Cambridge, graduating in 1883. In 1908 he became the first professor of genetics at Cambridge University, but he left in 1910 to take up the position of director at the newly established John Innes Horticultural Institution at Merton, Surrey, where he remained for the rest of his life. Bateson died in Merton on February 8, 1926.

Bateson became a controversial figure because he rejected the Darwinian idea that evolution is a steady, gradual process involving the slow accumulation of many small changes. Instead, he maintained that evolution occurs in occasional large spurts or jumps. He conducted breeding experiments to try to find support for his hypothesis, but found the evidence he was seeking when he read Mendel's paper on peas (see "Gregor Mendel and His Peas" on pages 144–147). Fearing that Mendel's paper might be lost for a second time, he translated it into English and then dedicated himself to publicizing and promoting Mendel's work.

DISCOVERY OF THE DOUBLE HELIX

In its April 25, 1953, issue, the science journal *Nature* carried a short article entitled "Molecular Structure of Nucleic Acids: a Structure for Deoxyribose Nucleic Acid." The authors, the American molecular biologist James D. Watson (1928–) and the English physicist Francis Crick (1916–2004), had discovered that the extremely large deoxyribonucleic acid (DNA) molecule formed a double helix. The

English biophysicist Rosalind Franklin (1920–58), working at King's College, University of London, with the New Zealand–born molecular biologist Maurice Wilkins (1916–2004) had crystallized DNA, then photographed it using X-rays. This revealed the symmetry of the molecule, and Watson and Crick, at the University of Cambridge, interpreted Franklin's results as showing that the molecule consisted of two helices, wound around one another and running in opposite directions. James Dewey Watson, Francis Harry Compton Crick, and Maurice Hugh Frederick Wilkins were jointly awarded the 1962 Nobel Prize in physiology or medicine for this discovery. Rosalind Franklin had died before the award was made, and Nobel rules forbid posthumous awards.

Biologists had long known that heredity involves the transmission of information from parents to their offspring. They had discovered that this information is contained in genes that are carried on chromosomes. Chemical analysis had revealed that nucleic acids—acids found in cell nuclei—played an essential part in the transmission of information and that DNA was the key ingredient. But how was the information conveyed?

Watson and Crick's article provided the answer, and that is why their discovery was so important. A single strand of the DNA molecule consists of a backbone made from units of deoxyribose, which is a sugar, linked by phosphate groups. A base attached to each deoxyribose group projects outward from one side of the backbone, and there are four bases: adenine, thymine, guanine, and cytosine. Bases will link to one another, but only in a very specific fashion: Links form only between adenine and thymine, and between guanine and cytosine. A single strand of DNA will therefore attract complementary bases and these will attach to deoxyribose units, forming a backbone, to produce a second strand of DNA linked to the first, with the two strands wound around each other. The illustration on page 153 shows the arrangement.

When the genetic instructions carried by DNA are to be read, *enzymes*—proteins that cause or accelerate chemical reactions in living cells without being altered themselves—first separate the strands of DNA. Molecules attach themselves to the bases, forming one type of ribonucleic acid (RNA); in RNA uracil replaces thymine. A sequence of three bases, called a *codon*, codes for a particular amino acid, and the list matching codons to the cor-responding amino acids constitutes the *universal genetic code* (see sidebar on page 154). This RNA then acts as a template for the synthesis of amino acids and the assembly of amino acids into protein molecules.

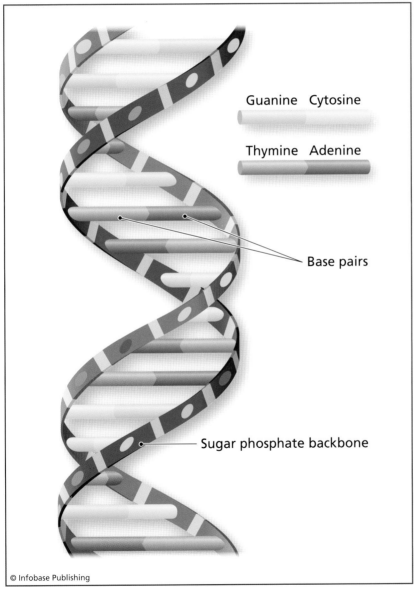

Guanine Cytosine

Thymine Adenine

Base pairs

Sugar phosphate backbone

© Infobase Publishing

A DNA (deoxyribonucleic acid) molecule consists of a backbone of sugar phosphate to which four bases are attached: adenine, thymine, guanine, and cytosine. The bases bond to bases on another backbone, adenine always bonding to thymine and guanine to cytosine. This forms a double strand that twists into a double-stranded helix.

Francis Crick realized that the sequence of steps by which instructions carried in the pattern of DNA bases are used to construct proteins must occur in only one direction. In 1958 he described the sequence, and it soon became known as the central dogma of molecular biology (see the sidebar on page 155).

Revealing the structure of DNA also revealed how evolution occurs. DNA is copied each time cells divide, but the copying process

THE UNIVERSAL GENETIC CODE

AMINO ACID	ABBREVIATION	CODONS
alanine	Ala	GCA, GCC, GCG, GGU
arginine	Arg	AGA, AGG, CGA, CGG, CGC, CGU
asparaginine	Asn	AAC, AAU
aspartic acid	Asp	GAC, GAU
cysteine	Cys	UGC, UGU
glutamic acid	Glu	GAA, GAG
glutamine	Gln	CAA, CAG
glycine	Gly	GGA, GGC, GGG, GGU
histidine	His	CAC, CAU
isoleucine	Ile	AUA, AUC, AUU
leucine	Leu	CUA, CUC, CUG, CUU, UUA, UUG
lysine	Lys	AAA, AAG
methionine	Met	AUG
phenylalanine	Phe	UUC, UUU
proline	Pro	CCA, CCC, CCG, CCU
serine	Ser	AGC, AGU, UCA, UCC, UCG, UCU
threonine	Thr	ACA, ACC, ACG, ACU
tryptophan	Trp	UGG
tyrosine	Tyr	UAC, UAU
valine	Val	UAC, UAU
stop codon		UAA, UAG, UGA

A = adenine; C = cytosine; G = guanine; U = uracil
(In RNA uracil takes the place of thymine)

THE CENTRAL DOGMA

The discovery of the structure of nucleic acids—deoxyribonucleic acid (DNA) and a number of ribonucleic acids (RNAs)—and the recognition that DNA encodes for amino acids, which join to form proteins, led to the description of the sequence of events involved. This sequence, first described in 1958 by Francis Crick, has come to be known as the central dogma of *molecular biology*—the study of the structure and functions of large molecules involved in biological processes, and especially DNA and RNAs. There are four steps in the sequence:

1. The two strands of DNA unwind with the help of enzymes; binding proteins attach to the single DNA strands to keep them straight; and other enzymes and primer molecules assemble a partner strand attached to each of the separate DNA strands. This is the process of replication.
2. DNA codes for the production of messenger RNA (mRNA). This is the process of transcription.
3. In eukaryotic cells, mRNA travels from the nucleus to the cell cytoplasm.
4. mRNA carries code to ribosomes, which follow its instructions to assemble proteins. This is the process of translation.

In addition, it is possible for information to pass from RNA to DNA in a process called reverse transcription. RNA can give rise to more RNA, in the process of RNA replication. Under some circumstances it is possible for DNA to be translated directly into protein.

is not perfect, and there are many genes; humans have about 23,000 genes made from about 3 billion pairs of bases. Mistakes happen. Sections of bases can be repeated, sections can be inserted or deleted, or one base may be substituted for another. In the case of Mendel's peas, for example, inserting a sequence of 800 base-pairs results in wrinkled rather than smooth peas. A change in the structure of a gene or chromosome is called a *mutation.*

GENOMES AND EVOLUTIONARY TREES

The full sequence of genes carried in almost every cell is known as the *genome* of the organism carrying it, and by the 1970s it was starting to become possible to identify and list entire genomes. Arranging the genes in their correct order to represent a genome is known as *sequencing.* The first full genome sequence was published in 1976. It was the genome of a virus and consisted of RNA rather than DNA, and it was 3,569 base pairs long. Whether a virus is a living organism is uncertain, but in 1995 the sequence of a bacterial genome was published. It was of *Haemophilus influenzae* (despite the name it does not cause influenza, although at one time it was thought to do so), and it comprised 1.83 million DNA base pairs. The first plant genome, of thale cress (*Arabidopsis thaliana*), with 157 million base pairs, was published in 2000.

It is tempting to imagine that because animals are more complex than plants and microorganisms their genomes are longer, but this is not the case. The largest genome known belongs to *Amoeba dubia* and consists of 670 billion base pairs, making it 200 times larger than the human genome. The first animal genome to be published, in 1998, was of the nematode *Caenorhabditis elegans*. It is 98 million base pairs long.

In 1991 the Human Genome Project began to assemble the full sequence of the human genome. This international project, undertaken by the National Center for Human Genome Research, was based at the U.S. National Institutes of Health and at first it was headed by James D. Watson and Victor Schmerkovich. Watson resigned in 1992, Francis Collins (1950–   ) took his place, and the institute's name was changed to the National Human Genome Research Institute. The Celera Corporation ran a parallel project. Both projects released a first draft of the human genome in 2000, and the National Center for Human Genome Research published what it defined as the full sequence late in 2003. It comprised 3.2 billion base pairs, but certain very repetitive regions of the genome that are difficult to isolate remained unsequenced.

The sequencing of genomes tells biologists nothing about what the genes do, but gene functions are also being revealed. Scientists now know a great deal about how genes translate into physical structures. Much less is known about the extent to which psychological traits are inherited. In the 19th century, however, many of Darwin's most enthusiastic followers jumped to the conclusion that people inherited their intelligence and personalities, and it seemed to them that these characteristics should be subject to natural selection in the same way as physical characteristics. That idea led to the growth of *eugenics*—a movement to improve the quality of human populations through selective breeding (see the sidebar on page 157). The eugenics movement flourished from the late 19th century until the 1940s, when horror at the way it had been interpreted and implemented in Nazi Germany brought the whole concept into disrepute.

The study of genomes has already affected the classification of organisms, building on work that was started by the German zoolo-

EUGENICS

Darwin's evolutionary theory is based on natural selection, according to which those individuals possessing traits that give them greater access to resources will produce more surviving offspring than individuals lacking those traits, and the offspring will inherit the advantageous traits. That implies that most offspring die before they are old enough to reproduce, it requiring parents to produce only two surviving offspring in order to maintain the population.

Natural selection applies only to heritable traits, but for a long time it was far from clear which traits were heritable and which were not, and in the 19th century some people supposed that such qualities as strength of character, honesty, and intelligence were inherited and, therefore, fixed at birth. No matter how carefully a child was nurtured or what educational opportunities were afforded to it, nothing could modify its genetic endowment. Along with the growth of this idea went the rapid advances in public hygiene and medicine that drastically reduced perinatal and infant mortality in North America and Europe. Infants were surviving that previously would have died and, supposedly, the surviving infants included many that natural selection would otherwise have removed.

Some scientists began to fear that medical advances might undermine natural selection and that this would lead inevitably to deterioration in the quality of human populations. In 1883, Sir Francis Galton (1822–1911), a cousin of Darwin and keen supporter of his theory, proposed a eugenics—literally well born—movement to remedy the situation. The eugenics movement sought to improve the human stock through selective breeding, which substituted for natural selection. It involved compulsory sterilization to prevent those with what were thought to be undesirable traits from having children. During the first decades of the 20th century, eugenic policies were adopted in many countries including the United States, reaching their most extreme form in Nazi Germany, where euthanasia was used to prevent those the Nazis categorized as undesirables from breeding.

Eugenics fell from favor, due mainly to shock at the Nazi application of it, but also because the underlying concept was fundamentally flawed. The categorization of some psychological and personality traits as desirable and others as undesirable is subjective and varies over time and from one culture to another. It is far from certain that such traits are inherited. If they are inherited, most individuals will be heterozygous for them, and recessive heterozygous traits are almost undetectable, so it is impossible to identify so-called undesirables. Every individual is heterozygous for many deleterious traits, so if a eugenic policy were to be implemented consistently, no one would be allowed to breed.

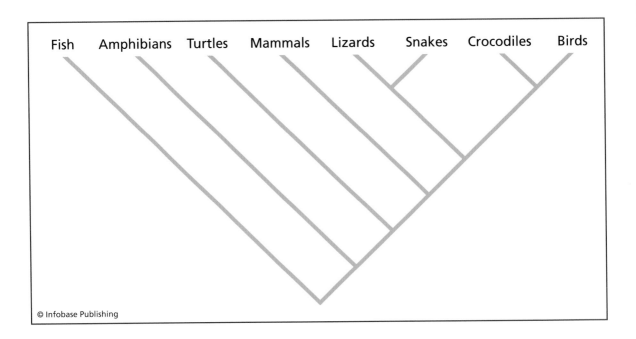

Fish   Amphibians   Turtles   Mammals   Lizards   Snakes   Crocodiles   Birds

© Infobase Publishing

A phylogeny shows the evolutionary relationships among organisms or, as in this example, groups of organisms. The diagram shows the evolutionary distance between the groups as a line leading from fish to birds, with each of the intermediate groups emerging in a sequence.

gist Emil Hans Willi Hennig (1913–76). In 1950 Willi Hennig published a book setting out a new approach to classification. The book was called *Grundzüge einer Theorie der phylogenetischen Systematik* (Basis of a theory of phylogenetic systematics). Hennig worked in Leipzig, then in the German Democratic Republic, his book was in German, and his ideas attracted little attention until 1966, when an English translation appeared, called *Phylogenetic Systematics*. This rapidly achieved an authoritative status that many zoologists have compared with the *Origin of Species*.

Hennig argued that classification must depict relationships, and relationships can be measured as the distance from a common ancestor. The pattern formed by lines of descent linking species to a common ancestor is called a *phylogeny*. The illustration above shows a phylogeny for all the vertebrates, from fish to birds.

A biologist working out the relationships between organisms lists the physical characteristics of the specimen and then compares them with those of related species. The comparison allows the scientist to distinguish between characteristics that have arisen recently in the specimen's evolutionary history and those that the specimen shares with other species. Wherever a new species emerges, the line on the diagram branches, producing two daughter species. This classification technique is known as *cladistics*, and it generates evolutionary trees that illustrate relationships very clearly. Cladistics aroused strong opposition at first from biologists who preferred more traditional methods for classifying organisms, but eventually it was accepted, and it is now the standard approach to taxonomy.

Animal Behavior

Although it is highly doubtful whether human personalities and tastes are acquired genetically, all animals inherit some of the behavior they exhibit. All animals, including humans, withdraw a limb if it touches something very hot and blink if an object approaches their eyes. These are automatic responses and, while evolutionary biologists and geneticists were decoding the genes governing the development and structure of bodies, other biologists were exploring the way animals behave.

This final chapter outlines, very briefly, some of the developments in studies of animal behavior. It begins with the observation, illustrated by many televised wildlife documentaries, that for most animals life is short and ends brutally. This links to ideas about the implications of natural selection for human society, which is sometimes called a "jungle." The chapter then proceeds to describe the work of some of the most respected animal behaviorists.

RED IN TOOTH AND CLAW

Hunters play a psychological game with their quarry. The hunter must keep upwind of the prey, which has an acute sense of smell. Hunters must move cautiously, because sudden movements will alert their prey. Alternatively, those that hunt by ambush must know where to set their traps. Hunters, be they human or nonhuman, must understand the habits of their prey, their degree of alertness, and how they respond to perceived danger. If they are to succeed—indeed, if they are to survive if hunting is their only means of obtaining food—hunters must understand animal behavior, at least as it applies to the animals they hunt.

People have always known this, of course, and in the 18th and 19th centuries, when there was growing interest in natural history, many people saw the relationship between animal species in very harsh terms. The natural world appeared to them as a bleak place ruled by a law of kill or be killed. Darwin's discovery and lucid explanation of natural selection seemed to reinforce this idea, but it did not initiate it. The picture of the natural world that was emerging, and to which Darwin contributed, was replacing the gentle and stable relationships established by divine law with a world of ruthless competition. The emerging picture also revealed that humans were not apart from the natural world. The poet Alfred, Lord Tennyson (1809–92), reflected on this in his long poem *In Memoriam A.H.H.*, which he wrote on the death of his friend Arthur Henry Hallam, at the age of 22. Tennyson completed the poem in 1849 and it was published in 1850, several years before Darwin published *Origin.* Canto 56 of *In Memoriam* contains the following lines.

> Who trusted God was love indeed
> And love Creation's final law—
> Tho' Nature, red in tooth and claw
> With ravine, shriek'd against his creed—

If the image of nature and the way animals behave toward one another was changing, so, too, were ideas about human society and the way it develops. During the second half of the 18th century agricultural practices were changing in much of Europe, and the agricultural changes were followed by changes in manufacturing. Goods that had formerly been made in small workshops or by people working at home were now produced in large factories on machines powered first by water and then by steam. This was the Industrial Revolution, and it accelerated through the first half of the 19th century. Socially, the rise of manufacturing and the resulting increase in trade promoted the growth and increasing prosperity of the middle class. To them, society seemed to be advancing rapidly on many fronts. Rising living standards brought better health, education, and opportunities for travel as well as material goods, and the tireless work of reformers promised to improve the lot of the poorest in society as well.

The firm belief in the inevitability of social and economic progress was accompanied by a belief in individualism. Individuals should be free to show enterprise and initiative, and society should reward talent and hard work. By implication, of course, those who lacked ambition or talent should be allowed to fail. It was an unforgiving doctrine, but the idea that human society was "red in tooth and claw" accorded with the prevailing view of the natural world. When Darwin's *On the Origin of Species* was published, in November 1859, the 1,250 copies sold out at once, and the book had to be reprinted several times. As readers absorbed Darwin's argument, many saw parallels between his description of natural selection and the human society around them. Their views later came to be known as social Darwinism (see the sidebar on page 163).

Applying Darwin's name to a political and economic philosophy suggests that it originated with Darwin. This is not so, and in fact Social Darwinism owes little to Darwin's work. The "survival of the fittest"—an expression coined by Herbert Spencer and adopted by Darwin in later editions of his work—implies only that individuals better adapted to their environment—the fittest—produce more off-spring and the offspring inherit whatever characteristic it was that made their parents fit. It does not imply that the fit oppress, bully, or cheat the less fit.

SIR JOHN LUBBOCK AND THE SOCIAL INSECTS

In the 19th century science had not yet become professionalized. It was possible for private individuals, provided they had the educa-tion and economic means, to conduct research that made valuable contributions to the accumulating body of scientific understanding. Indeed, in this sense most 19th century scientists—including Dar-win—were amateurs.

One of the first biologists to make detailed studies of animal behavior was a talented amateur. Sir John Lubbock (1803–65) was a successful banker, a popular author, and a member of parliament as well as a geologist, archaeologist, anthropologist, and botanist. At one time or another, Lubbock was president of 25 learned societies, and he received honorary degrees from the universities of Oxford,

Cambridge, Edinburgh, Dublin, St. Andrews, and Würzburg. As a politician, his most memorable achievement was the law, passed in 1871, that established bank holidays; at one time these were known

SOCIAL DARWINISM

In the 19th century European economies grew increasingly dependent on large-scale manufacturing. As they did so, the wealth that industrialization generated, and the rate at which new scientific discoveries were made and inventions appeared, produced a strong and optimistic belief in progress. Intellectuals could see very clearly the achievements that were possible when enterprising individuals seized opportunities to convert natural resources into goods. They could also see that those who lacked enterprise or talent fell by the wayside. Years before Darwin published his conclusions regarding evolution, some theorists were coming to see society as marching toward ever-greater perfection and rewarding initiative, energy, enthusiasm, and hard work. This seemed to chime with Darwin's explanation of natural selection.

The English philosopher Herbert Spencer (1820–1903) was the individual most closely associated with this sociological outlook. Spencer believed that evolution was a progress from a condition when all living organisms were identical, through their steady diversification, until life reached a state of perfection in which every individual was unique. This progress applied equally to human society, which began as a homogeneous mass, and in time divided into discrete groups such as classes, and ended when a perfect society was composed of perfect individuals perfectly adapted to their social life. Competition was the force driving progress, and competition occurred at every level—among individuals, groups in society, and nations. Spencer was not alone. The English philosopher Thomas Hobbes (1588–1679) had expressed somewhat similar views and Thomas Malthus (1766–1834) proposed that population size must inevitably be constrained by the availability of resources, implying strong competition for those resources.

Spencer first described his theory in an essay entitled "Progress: Its Law and Cause," that was published in 1857 in the philosophical journal *Westminster Review*. This was two years before the publication of *Origin of Species*. Spencer's ideas seemed similar to Darwin's description of natural selection, but Spencer did not consider natural selection to be socially significant. His theory was based on a version of Lamarckism: Individuals improve themselves by their own talent and efforts, their children inherit the gains their parents made and build on them, and society benefits as a whole.

In 1879 an article in the American monthly magazine *Popular Science* described this social theory as social Darwinism, and the following year *Le darwinisme social* was the title of an anarchist pamphlet published in Paris. Spencer never used this term himself. The expression remained obscure until 1944, when the American historian Richard Hofstadter (1916–70) popularized it in his book *Social Darwinism in American Thought, 1860–1915*. Since then social Darwinism has usually been applied in a pejorative sense, and it is often associated with far-right political philosophies.

as Saint Lubbock days. He made several contributions to archaeology, and it was Lubbock who coined the terms Neolithic and Paleolithic to describe the New and Old Stone Ages. As an author, most of his 27 books were about science, and many of them were translated into other languages.

In 1882 Lubbock published *Ants, Bees, and Wasps: Observations on the Habits of the Social Insects.* This book was a landmark, in that Lubbock described how he had constructed a glass-sided nest that allowed him to study the insects. He marked individuals with tiny dots of paint so he could track their movements and behavior, and he used mazes and obstacles to test the intelligence of ants. One of his most interesting techniques was to follow an ant's movement with a pencil, producing a tracing. He included many of these tracings in his book, each one with an explanatory caption. Lubbock continued his observations, mainly of ants, for many years. His descriptions were precise, detailed, and quantified, and he treated his data statistically. The book became very famous, attracting the attention of the humorous magazine *Punch,* which printed a cartoon showing Lubbock's head with the body of a bee, surrounded by flowers. The caption read:

> Sir John Lubbock, m.p., f.r.s.
> How doth the Banking Busy Bee
> Improve his shining Hours?
> By studying on Bank Holidays
> Strange insects and Wild Flowers!

John Lubbock was born in London on April 30, 1834. His father, Sir John William Lubbock (1803–65), was a successful banker, mathematician, and astronomer. Lubbock was educated at Eton College, then went to work in his father's bank, becoming a partner in 1856; Lubbock's bank later amalgamated with Coutts & Co. He was elected the Liberal Party member of parliament for Maidstone, Kent, in 1870 and 1874, and in 1880 he was elected M.P. for the University of London (at that time the major universities had their own M.P.s). While he was an M.P., Lubbock lived at Downe, where his neighbor living in Down House was Charles Darwin. The two became friends, and Darwin encouraged Lubbock's interest in natural history. Lubbock later helped to arrange Darwin's burial in Westminster Abbey. Lubbock was made a privy councillor in 1890 (the privy council is a body that advises the sovereign), and in 1890 he was made a peer, becoming the first Baron Avebury. He died on May 28, 1913.

MAX VERWORN, JACQUES LOEB, AND RESPONSES TO STIMULI

During the final decades of the 19th century certain biologists began to explore the basis of all behavior. Simple observation allowed zoologists to document the way animals behave in particular situations, but at what level does behavior originate? Do individual cells exhibit behavior, and, if they do, how is their behavior related to the behavior of the animals of which they are part? Phrasing the question a different way, at what point does a purely chemical process become biological?

This was a subject that the German physiologist Max Verworn (1863–1921) explored. Verworn studied the processes that take place in the cells of the nervous system, muscle fibers, and sensory organs. These processes combine to put the cells in a particular state or condition, and Verworn introduced the term *conditionalism* to describe this.

Max Verworn was born in Berlin on November 4, 1863. He studied medicine, zoology, and botany in Berlin and Jena, and in 1895 he became a professor at the University of Jena. In 1901 be was appointed professor of physiology at the Institute of Physiology in Göttingen, and in 1910 he was made a professor at the University of Bonn. He died in Berlin on November 23, 1921.

The German-American physiologist Jacques Loeb (1859–1924) studied *tropism*—the directional response of an organism to a stimulus. In 1886 Loeb obtained a post as assistant to Adolf Fick (1829–1901), professor of physiology at the University of Würzburg, where he became friendly with Julius von Sachs (1832–97), a professor of botany. Von Sachs was investigating tropism in plants, and Loeb wondered whether animals responded in the same way. When a seed germinates, for instance, a shoot grows toward a source of light and a rootlet grows downward in response to gravity. When the plant is established, roots also grow to the sides, following chemical stimuli as they search for moisture. Single cells that are free to move also move in response to light, gravity, chemical gradients, magnetic fields, and other simple stimuli. Loeb began by studying insects and discovered that they respond to the stimulus of light, in some cases moving toward light in preference to food. He also found that insects responded to gravity. Loeb came to believe that all animal behavior can be reduced to tropisms. He began to experiment in psychology, and in 1888 he wrote the following in a letter to the American philosopher William James (1842–1910):

> Whatever appears to us as innovations, sensations, psychic phenomena, as such are called, I seek to conceive through reducing them—in the sense of modern physics—to the molecular or atomic structure of protoplasm, which acts in a way that is similar to (for example) the molecular structure of the parts of an optically active crystal.

Loeb had begun by seeking to discover whether individuals possess free will. In the end he concluded that they do not. He had been convinced by the results of his research that all animals, including humans, were machines, albeit highly complex machines. Behavior of all kinds amounts to automatic responses to stimuli. Loeb's other major discovery was induced parthenogenesis. He found that by altering the chemical composition of seawater he could stimulate sea urchin eggs to begin to develop into embryos without fertilization.

Loeb was born as Isaak Loeb on April 7, 1859, in Mayen, in the Rhineland-Palatinate (Rheinland-Pfalz) in southwestern Germany. His father, Benedict, was an importer. Loeb's mother, Barbara Isay, died in 1873, and Benedict died in 1876. In 1875 Isaak had been sent to Berlin, where he worked for a time in his uncle's bank before continuing his education. He enrolled at the University of Berlin in 1880 to study medicine, and in that year he changed his first name to Jacques. After one semester he transferred to the University of Munich and then to Strasbourg. He received his medical degree at Strasbourg in 1884. He returned to Berlin in 1885 to work at the Agricultural College and went on to work at the universities of Würzburg (1886–88) and Strasbourg (1888–90), at Kiel in 1888, and at the Naples biological station (1889–91).

In 1890 Loeb met Anne Leonard, an American who had just received her Ph.D. from the University of Zürich. The couple married in October of that year and went on to have three children. In 1891 the family moved to the United States, where Loeb became a professor at Bryn Mawr College in Pennsylvania. He left Bryn Mawr in January 1892 and moved to the University of Chicago, becoming a professor of physiology in 1899. He remained there until 1902, when he moved to the University of California, Berkeley, where he stayed until 1910, when he became head of experimental biology at the Rockefeller Institute for Medical Research in New York City. Loeb remained in this position until his death on February 11, 1924, while on vacation in Bermuda. Loeb was recommended for a Nobel Prize in 1910, and in the same year he was elected to the National Academy of Sciences.

C. LLOYD MORGAN AND HIS CANON

Loeb's mechanistic view of animal behavior was highly controversial, and in the 1890s a horse called Clever Hans (Der Kluge Hans) seemed to disprove it. Clever Hans belonged to Wilhelm von Osten, a high school mathematics teacher and amateur horse trainer. Von Osten had taught Hans to add, subtract, multiply, divide, calculate with fractions, tell the time, and distinguish musical notes. Hans was also able to read, spell, and understand German. He communicated by tapping one hoof, with the number of taps indicating his answer, and he would reply to questions that were either spoken or written down. Although Hans could read, obviously Von Osten read the written questions out loud, so the audience could know whether Hans answered them correctly. In 1891 Hans began giving public performances, accompanied by von Osten, and his fame quickly spread.

Hans appeared to possess phenomenal powers. Various "experts" tested him and could find no sign of cheating, one of them asserting that Hans had the intellectual capacity of a 14-year-old boy. Professor Carl Stumpf of the Berlin Psychological Institute examined the horse and could find no fraud. Von Osten was an honest man, and the horse really could perform the feats the growing crowds enjoyed.

Not everyone was convinced, however. One of the skeptics was Oskar Pfungst, one of Stumpf's students. In 1907 Pfungst devised a series of more subtle tests that were conducted inside a large tent. The tests involved a large number of different questions from different questioners. Some of the questioners knew the answers to the questions they asked, others did not. The questioners stood at varying distances from Hans, and in some trials Hans wore blinkers.

Pfungst found that the farther away the questioner was, the more likely it was that Hans would give an incorrect answer; the horse needed to be able to see the questioner clearly. He also noted that Hans gave correct answers only if the questioner knew the answer. Pfungst concluded that Hans was, indeed, extremely clever—but not at mathematics or language. As his numbers of taps approached the correct answer, growing tension that was visible in the questioner's body or face told Hans when to stop. The questioner was not aware of giving clues, and neither was Wilhelm von Osten, but that is what was happening. Nowadays, when an animal makes responses that are influenced by unconscious signals from a questioner it is called the *Clever Hans effect*.

The case of Clever Hans demonstrated just how easy it was for even experienced scientists to imagine they are observing intelligent behavior in a nonhuman animal. After all, everyone knows or has heard of the dog that "understands everything I say to him." This problem led the English psychologist C. Lloyd Morgan in 1894 to issue a piece of advice that every student of animal behavior must learn. It has come to be known as *Morgan's canon,* and in his own words it states: In no case may we interpret an action as the outcome of a higher mental faculty, if it can be interpreted as the exercise of one which stands lower in the psychological scale. Scientifically, this is a version of the principle of parsimony often attributed to the Franciscan friar William of Occam (or Ockham, ca. 1285–ca. 1349). Also known as Occam's razor, it states that one should not increase, beyond what is necessary, the number of entities required to explain anything. It is called a "razor" because it involves shaving away all unnecessary explanations. In fact, similar statements were made earlier, by Alhasen (965–ca. 1040), Moses Maimonides (1135–1204), and John Duns Scotus (ca. 1266–1308).

Conway Lloyd Morgan was born in London on February 6, 1852. He studied at the Royal School of Mines (now part of Imperial College, London), partly under T. H. Huxley (1825–95), who was professor of natural history, and at University College, Bristol, receiving doctorates in science and law. After graduating, he spent five years teaching at the Diocesan College in Rondesbosch, South Africa. In 1884 Morgan returned to Bristol as a professor of geology and zoology. In 1901 he became the college's first professor of psychology and education. Morgan became principal of University

College in 1891. When University College became the University of Bristol in 1909, Morgan became its first vice-chancellor, but in 1910 he resigned that position to become professor of psychology and ethics, a post he held until 1919, when he retired. Morgan died in Hastings on March 6, 1936.

IVAN PAVLOV AND CONDITIONED RESPONSES

Pavlov's dogs are among the most famous of all experimental animals. Millions of people who know nothing of physiological or behavioral research have heard of the dogs that Pavlov taught to salivate at the sound of a bell. The response was automatic, and some people might describe it as instinctive. Instinct and the rather similar drive have fallen from use among students of animal behavior, however, partly because Pavlov demonstrated how behavior can be modified (see the sidebar below).

Pavlov was a physiologist who in 1890 had helped establish a department of physiology at the Institute of Experimental Medicine at the University of St. Petersburg. He was the department's first director. His own research was into the digestive processes, and he devised a technique for observing the way internal organs function without having to perform vivisection. Between 1891 and 1900 he made many discoveries that were highly relevant to the practice of medicine. He found it was mainly the nervous system that regulated digestion, and that digestive organs reacted automatically to stimuli.

INSTINCT AND DRIVE

If an animal always responds to a particular stimulus with the same, fixed behavior it is popularly said to be acting by instinct. Instinctive behavior is inherited genetically. The term is little used by biologists, because it takes no account of environmental conditions that may influence behavior and behavioral responses, and because behavior that was once thought to be instinctive in fact results from a variety of motivations.

Drive was used to describe the strong motivation for behavior involving physical action. Biologists no longer use the term. Drive suggests energy, in the physical sense, and no such energy is involved in psychological processes. Scientists also found that if every behavior resulted from a drive, they faced an endless number of drives, with no explanation for any of them. This rendered the concept useless.

He became especially interested in salivation. Pavlov knew that a dog will salivate when it smells food (he used meat powder) but noticed that after a time his dogs began to salivate when they saw the technician who normally fed them. He called this salivation psychic secretions, and that is what he set out to investigate with his famous experiments with dogs.

The illustration below shows the experimental arrangement. This was prepared in a quiet room where, so far as possible, there was nothing to distract the dog. Pavlov attached a hungry dog to a frame so that it was restrained without being made uncomfortable. The dog faced a panel and could be exposed to sounds, lights, and other stimuli. A tube passed through the dog's cheek to collect saliva and carry it outside the animal, where the amount was measured and recorded, and a technician could puff a measured amount of meat powder into the dog's mouth. Pavlov found that when meat powder was blown into the dog's mouth, the amount of saliva it secreted

Pavlov placed dogs under gentle restraint facing a panel. This setup allowed him to present the dog with a variety of stimuli and measure the dog's response.

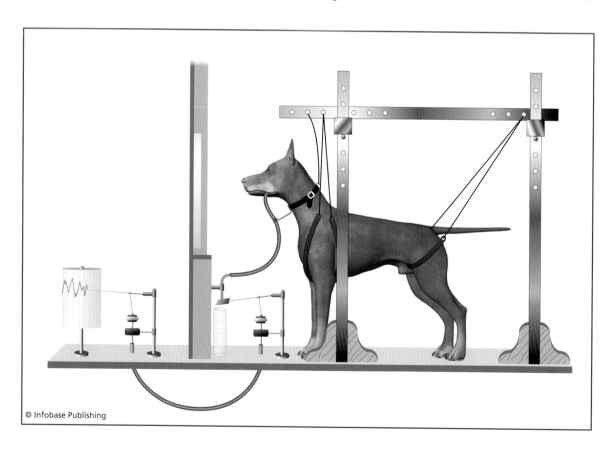

was proportional to the quantity of meat powder. He then provided a sound stimulus just before administering the meat powder. The dog clearly heard the sound—it moved its ears—but at first it did not salivate. After Pavlov had repeated this procedure about six times, however, the dog began to salivate on hearing the sound and before receiving the meat powder. Finally, the sound stimulus triggered the release of as much saliva as did a measure of meat powder.

Pavlov called the meat powder an *unconditional stimulus,* because its significance was obvious to the dog. He called the sound—he used a metronome as well as bells—a *conditioned stimulus* because the dog had to learn to associate it with the unconditional stimulus. In later experiments Pavlov found that any stimulus would produce this effect, provided the stimulus was not one to which the dog ordinarily had a strong response. Once a dog was conditioned to salivate in response to a particular type and pitch of sound, it would also respond to other sounds that were similar, but the strength of the response diminished as the sound became more different. The effect Pavlov had discovered is known as *classical conditioning* or *Pavlovian conditioning.*

Ivan Petrovich Pavlov was born on September 14, 1849, at Ryazan, southeast of Moscow, the son of a village priest. He attended the church school and then a theological seminary, both in Ryazan. At first he intended to become a priest, but then changed his mind and in 1870 he enrolled at the University of St. Petersburg to study natural science. He graduated in 1875 and qualified as a surgeon in 1879. He became director of the physiological laboratory at a famous clinic, where he continued his physiological research, and in 1883 he successfully submitted his doctoral thesis on "The centrifugal nerves of the heart." Pavlov was appointed professor of pharmacology at the Military Medical Academy in 1890, and in 1895 he became professor of physiology, a position he held until 1925. He held his post at the Institute of Experimental Medicine for 45 years, until his death. Pavlov was awarded the 1904 Nobel Prize in physiology or medicine for his work on digestion. He died in Leningrad (now St. Petersburg) on February 27, 1936. His laboratory is now a museum.

KONRAD LORENZ AND IMPRINTING

Ethologists—biologists who study animal behavior, especially of wild animals—no longer use the term *instinct,* but in the 1930s they still did so. In 1936, at a symposium held in Leiden, the Netherlands, to discuss instinct, the Austrian ethologist Konrad Lorenz (1903–89) met for the first time a Dutch colleague, Niko Tinbergen (see "Niko Tinbergen and Defending Territory" on pages 174–176). The two became lifelong friends, but at the symposium it was geese that interested them most. In 1937 Tinbergen spent several months at Lorenz's home collaborating with him on the egg-rolling response of graylag geese (*Anser anser*). They found that if they removed an egg from a goose that was incubating it and placed the egg about 3 feet (1 m) away, the goose would invariably extend its neck, tuck the egg beneath its bill, and gently roll the egg back into the nest.

A baby quickly learns to recognize the face of the person it sees most frequently, usually its mother. This process of rapid learning is called *imprinting*, and it occurs in species that are not able to fend for themselves from the time they are born or hatch. The English biologist Douglas Spalding (1841–77) observed this behavior in chickens and was the first scientist to report it. The German biologist Oskar Heinroth (1871–1945) also studied it, and Heinroth's student Konrad Lorenz studied it in great detail and popularized it in books and TV programs. Lorenz incubated goose eggs in an incubator and found that after they hatched there was a period of 13 to 16 hours, which he called the critical period, during which the goslings would imprint onto the first moving object they saw. Once imprinted, the goslings would follow that object wherever it went. He arranged matters so that he was the object onto which the goslings imprinted, and there are many movies and photographs of him walking along followed by a family of young geese. Of course, having made the birds imprint on him, Lorenz had a responsibility to care for them. He worked most with graylag geese. The photograph on page 173 shows Lorenz crossing a field with food for his graylag family.

Konrad Lorenz is sometimes called the father of ethology. In 1969 he was the first recipient of the Prix Mondial Cino Del Luca, an international literary award established in France by Simone Del Luca (1912–2004). Lorenz shared the 1973 Nobel Prize in physiology or medicine with Niko Tinbergen and Karl von Frisch (see "Karl von Frisch and the Dances of the Honeybee" on pages 176–179).

Konrad Zacharias Lorenz was born in Altenberg, near Vienna, Austria-Hungary, on November 7, 1903. From his earliest years he had a love of animals, which his parents indulged. His father, Alfred,

was a wealthy orthopedic surgeon. Konrad wanted to become a zoologist and paleontologist, but at his father's request he studied medicine, enrolling in 1922 for a premedical course at Columbia University, New York City, and returning in 1923 to study medicine at the University of Vienna. He qualified as a doctor in 1928 and taught at the university until 1935. He received his Ph.D. in zoology in 1933. In 1940 he became a professor of comparative psychology at the University of Königsberg, Germany. In 1941 Lorenz was drafted into the German army as a doctor, although he had never practiced medicine. He was captured by the Red Army in 1942 and remained a prisoner of war in the Soviet Union until 1948, continuing to work as a doctor. On his return to Austria, Lorenz had no job, but friends found him some lecturing and the Austrian Academy of Sciences established a small experimental station for him at Altenberg with funds provided by the English author J. B. Priestley. This became the Konrad Lorenz Institute for Ethology, and it is still conducting research. In 1950 the

The graylag geese following Konrad Lorenz believe he is their mother. *(Nina Leen/Time Life Pictures/Getty Images)*

Max Planck Society established the Lorenz Institute for Behavioral Physiology at Buldern, Germany, with Lorenz as its director. Lorenz retired in 1973, but continued to research and write from his home, where he died on February 27, 1989.

NIKO TINBERGEN AND DEFENDING TERRITORY

In 1953 the Dutch-British ethologist Niko Tinbergen (1907–88) performed a set of experiments with three-spined sticklebacks (*Gasterosteus aculeatus*)—fish about 2–4 inches (5–10 cm) long that live in fresh or brackish water. In winter these fish live in schools, but as the mating season approaches in spring the males develop red bellies and become aggressive, establishing territories that they defend against rival males. Sticklebacks use their territories solely for reproduction, the male building a nest and enticing females to enter and lay eggs, which the male fertilizes and then guards. These fish are not territorial outside the mating season.

A male will drive away any other male that enters its territory, but no matter how big and strong it may be, a male that enters the territory of another male always retreats. Tinbergen studied this behavior. First he released two male sticklebacks, *a* and *b*, into an aquarium and allowed them to establish territories. Then he caught the two fish and placed each in its own glass tube. When he placed both tubes inside *a*'s territory, *a* tried to attack *b* through the glass and *b* tried to escape. When he placed both tubes inside *b*'s territory, *b* tried to attack *a* and *a* tried to flee. Tinbergen's experiment showed that in territorial disputes the intruder usually loses. But why does this happen? It cannot be because the occupant of a territory must be strong and combative in order to establish and hold a territory in the first place, because Tinbergen's sticklebacks were evenly matched and both occupied territories. Tinbergen concluded that among territorial species individuals accept and acknowledge boundaries and avoid the risk of injury by settling border disputes peaceably; the occupant declares its claim and the intruder retreats. It is not strength that decides the winner, but the space each contestant occupies. Tinbergen also conducted detailed studies of the courtship behavior of sticklebacks. He described this work, including his investigation of territoriality, in *The Study of Instinct*, a book published in 1951 that became a classic among textbooks.

The stickleback experiment was simple, but it produced clear results. Some of his earlier experiments were equally simple. In one investigation, for instance, Tinbergen wished to discover how wasps that live in burrows find their own burrows when returning to them. He used arrangements of pinecones to establish that the wasps recognize landmarks close to the burrow entrance. He also studied communication among herring gulls (*Larus argentatus*), which are also territorial during the breeding season. He described that work in his book *The Herring Gull's World*, published in 1953. The illustration below shows Niko Tinbergen at about this time.

Tinbergen helped establish the study of animal behavior as an observational and experimental scientific discipline. His contribution was rewarded in 1973, when he received a share of the Nobel Prize in physiology or medicine (see "Konrad Lorenz and Imprinting" on pages 171–174 and "Karl von Frisch and the Dances of the Honeybee" on pages 176–179). Tinbergen was not the only Nobel laureate in his family; his elder brother Jan won the 1969 Nobel Prize for economics.

Nikolaas (Niko) Tinbergen was born in The Hague on April 15, 1907, the third of the five children of Dirk Cornelius Tinbergen, a schoolteacher, and Jeanette van Eek. Fascinated by wildlife from an early age, after leaving high school he worked for a time at the Rossitten bird observatory—then in Germany, now in Russia, and known as the Biological Station Rybachy of the Zoological Institute, Russian Academy of Sciences—before enrolling at the State University of Leiden to study biology. He received his doctorate in 1932 for his thesis on insect behavior. Soon afterward he married Elisabeth Rutter, and the couple joined an expedition to Greenland as part of the Netherlands's contribution to the International Polar Year 1932–33. He described their experiences in *Curious Naturalists*, a book published in 1958. On their return, Tinbergen obtained a post as lecturer at Leiden University. In 1938 he spent four months in the United States. When German forces occupied the Netherlands in World War II, Tinbergen protested at the university's decision to dismiss three faculty

Nikolaas Tinbergen, the Dutch zoologist and ethologist *(Science Photo Library)*

members because they were Jewish. As a result, he was held in a German prison camp from 1942 until 1944.

After the war Tinbergen returned to Leiden, and in 1947 he was made professor of experimental zoology. He held this position until 1949, when he took up an appointment as professor of zoology at the University of Oxford, where he established a school of animal behavior. Tinbergen remained at Oxford until he retired in 1974. He became a British citizen in 1955. In addition to his academic work, Niko Tinbergen wrote books about wildlife for children and made many natural history films. In the years following his retirement, Tinbergen and his wife studied behavioral disorders in humans and autism in children. Tinbergen died at his home in Oxford on December 21, 1988.

KARL VON FRISCH AND THE DANCES OF THE HONEYBEE

Naturalists had known for centuries that when social insects locate a source of food they congregate at it. This can happen only if the insect that finds the food source is able to inform other members of the colony of its existence and location. It follows that social insects must possess some means of communication. Sir John Lubbock (see "Sir John Lubbock and the Social Insects" on pages 162–165) had studied ants in careful detail and could see that the ants followed a trail left by the ant that found the food. He could see that this was happening, but it was many years before biologists discovered that the first ant leaves a scent trail that the others follow.

A chemical substance released by an animal into the environment that induces a behavioral response in another animal of the same species is called a *pheromone*, a term that was introduced in 1959 by P. Karlson and M. Lüscher in an article in the journal *Nature*. Pheromones can work as scent trails for insects such as ants that walk, and many flying insects release pheromones into the air to attract mates; attractant scents drift with the wind and produce a gradient of intensifying scent that guides the insect following it toward the source. It is difficult to see how pheromones could guide bees to a food source, however, yet for centuries beekeepers have known that this happens. The scientist who discovered the way that bees communicate was the Austrian-born German ethologist Karl von Frisch (1886–1982).

Von Frisch was awarded a share of the 1973 Nobel Prize in physiology or medicine for his work (see "Konrad Lorenz and Imprinting" on pages 171–174 and "Niko Tinbergen and Defending Territory" on pages 174–176).

Von Frisch constructed hives that allowed him to observe the behavior of the bees inside, and he marked individual bees so he could follow their movements. What he discovered was that when a foraging bee finds a source of food, on its return to the hive it per-

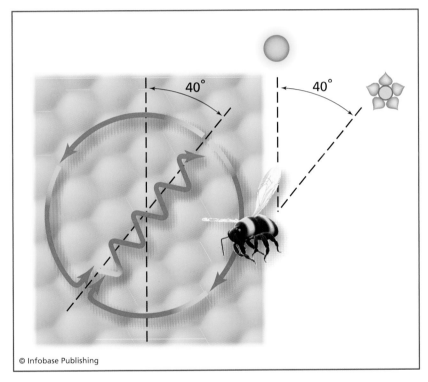

© Infobase Publishing

A bee returning to the hive informs other workers of the location and distance of a food source it has discovered. It performs a waggle dance on the vertical comb, in the total darkness of the hive. The dance has a figure-of-eight pattern, with a zigzag path (the waggle) between the two halves of the eight. The angle the waggle path makes with the vertical is the same as the angle between the Sun and the direction of the food (in this example 40°). The length of the waggle run and the number of direction changes indicate the distance. If the bee performs the dance outside the hive in daylight, the dance is horizontal and the waggle path points in the direction of the food source.

forms a dance. If it dances in a circle, called the *round dance,* the food is close to the nest. If it dances along a figure-of-eight path and follows a zigzag path between the two halves of the eight, the zigzag section indicates the direction of the food source and the distance to it. This is called a *waggle dance.* The illustration on page 177 explains its meaning.

The German zoologist Martin Lindauer (1918–2008), a student of von Frisch who later became a professor of zoology at the University of Frankfurt, showed that there is more to the dance language than even von Frisch had suspected. Lindauer found that the speed with which workers in the hive unloaded the nectar from a returning forager indicated whether or not nectar was in short supply. The greater the hive's need for nectar, the more likely it was that the returning bee would dance to direct other foragers toward a food source. Lindauer and von Frisch also collaborated in research demonstrating that bees are able to calculate a compass direction from their knowledge of time of day and the way the Sun moves across the sky.

Karl von Frisch experimenting at his home at Munich. He is using yellow cards to attract bees in order to test their color vision. *(Nina Leen/Time Life Pictures/Getty Images)*

Karl von Frisch was born in Vienna, Austria-Hungary, on November 20, 1886, the son of a university professor. He was educated at the Schottengymnasium (high school) in Vienna and enrolled at the University of Vienna to study medicine, but transferred to the University of Munich, where he studied zoology. He also studied marine biology at the Trieste Biological Institute for Marine Research. He received his Ph.D. degree from Munich in 1910 for his research into color perception and light adaptation in minnows. After graduating he remained at Munich as a research assistant, and in 1912 he was promoted to lecturer in zoology and comparative anatomy. During World War I his poor eyesight meant he was not drafted into the Germany army, but instead he worked at a Red Cross hospital. He was a professor and director of the Zoological Institution at Rostock, Germany, from 1921 to 1923, when he moved to a similar position at the University of Breslau. Von Frisch returned to the University of Munich in 1925 and established its Zoological Institution.

This was destroyed during World War II, however, and in 1946 von Frisch moved to the University of Graz, in Austria. He remained there until 1950, when he returned to Munich, where he stayed until he retired in 1958. The photograph on page 178 shows him late in his career, experimenting on color vision in bees at his home. Von Frisch died at Munich on June 12, 1982.

CAMERAS, RADIO COLLARS, AND EVENT RECORDERS—MODERN BEHAVIOR STUDIES

The study of animal behavior is based on observation, and observing animals in their natural habitats is time-consuming and calls for great patience. Modern ethologists have more powerful equipment than their predecessors, and, although this does not reduce the number of hours they must spend outdoors waiting for an animal to appear, it greatly increases the amount of information they are able to obtain, allowing them to use their time more productively. The new instruments allow scientists to track animals without disturbing them, watch them from a greater distance and in total darkness, record their presence automatically in the absence of the researcher, and record their observed behavior more conveniently.

Digital cameras will now operate in light levels lower than those possible with film, and their pictures are easier to process. The researcher does not always need to be present, because cameras can be activated automatically when an animal interrupts an infrared beam. The most important advance in this area, however, is with the use of CCTV (closed-circuit television) cameras to monitor animal behavior under remote control. Night-vision cameras and CCTV recorders allow animals to be observed in total darkness. This is an especially valuable tool, since many animals are nocturnal.

A *radio collar* is a collar with a small radio transmitter attached to it. An animal is captured, tranquilized if necessary, and fitted with the collar. Depending on its power, the device transmits a steady signal over a range up to about 500 yards (460 m) that can be detected by a receiver with a directional aerial. The researcher can follow the animal subject, reliably tracking its movements but always remaining far enough away for the subject not to be disturbed. Transmitters are now so small that they can be attached to very small animals. Animals that are too small to wear a collar can have a transmitter glued to them. Collars and glue are designed to deteriorate, so that after a certain length of time they fail and fall from the animal. Radio telemetry has greatly improved the capacity for monitoring animal movements. Individuals are fitted with transmitters that emit signals detected by orbiting satellites.

Birds are most often tracked by fitting rings to their legs. This is an ancient technology. The Roman army sent messages in this way around 210 B.C.E., and medieval falconers attached metal plates to their falcons for identification. Scientific bird ringing to track bird movements began in 1899 when a Danish schoolteacher fitted zinc rings to European starlings (*Sturnus vulgaris*). The first organized bird-ringing scheme was established in 1903 at the Rossitten bird sanctuary on the German (now Russian) Baltic coast. This was the world's first bird sanctuary; today it is known as the Biological Station Rybachy of the Zoological Institute, Russian Academy of Sciences.

Bird rings are stamped with information identifying the station where they were fitted. Researchers capture migratory birds, usually by netting them, and can notify the place where the rings were fitted. Radio telemetry is now replacing ringing. Careful studies have shown that animals quickly accept the equipment attached to them, and it has no effect on their behavior.

Traditionally, a behavioral researcher records observations against a checklist on printed forms. With practice it is possible to become so familiar with the layout of the form that the observer barely needs to glance down to tick a column, thus minimizing the time spent looking away from the subject. An *event recorder* is a portable electronic device that performs the same task. It has a keyboard that a researcher uses to type in coded observations of behavior. The keyboard may be arranged like a standard QWERTY keyboard or be more specialized, and the device can be connected to a computer, which will display the results. In some types the software includes statistical programs that process the data. There are also event recorders that work by voice recognition, for use in situations where the researcher's voice will not disturb the subjects.

Technological advances are enabling behavioral researchers to obtain much more information about many more species than has ever before been possible. This increasing power to observe and monitor wildlife is applied in conservation programs, allowing scientists to identify species that are at risk and the nature of the threats to them.

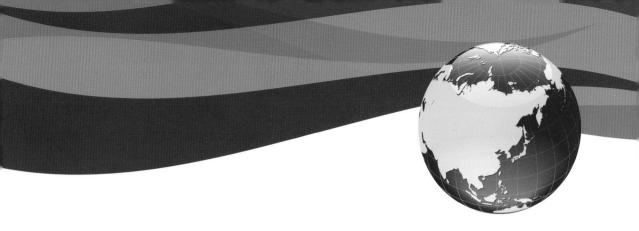

Conclusion

Zoological and behavioral research continues apace. Changes in land use are reducing areas of natural habitat in many parts of the world, and in some regions pollution and hunting are taking their toll on animal populations. In its *IUCN Red List of Threatened Species*, the International Union for Conservation of Nature (IUCN) estimates that of all the known animal species whose conservation status has been evaluated, 27 percent are threatened with extinction, although the level of threat varies. If the rate of animal extinctions is to be reduced, scientists engaged in conservation must have access to detailed information about animal species and populations. Continuing animal research is essential.

People today admire wild animals and support organizations dedicated to their protection. As this book has shown, however, attitudes to animals have changed throughout history, and it is only recently that scientists have been in a position to draw attention to the threats that many now face. In past centuries wild animals have been used to demonstrate political power, and they still feature in many flags and coats of arms. Originally lions, dragons, bears, eagles, and other strong and dangerous beasts were chosen to impress friends and intimidate foes. Modern organizations often prefer smaller and gentler symbols—but while conveying the core interests or activities of a body these also form part of coats of arms, devices historically associated with authority and power. The symbolism is subtler, but the meaning is much the same. Animals, then, are often symbols, and for many years they were also used for moral instruction. Every animal had a moral tale to tell—and some of that imagery survives.

Meanwhile, natural historians were exploring the origins of animals. This study led to investigations of the evolution and evolutionary history of the animals living today. Darwin proposed an evolutionary mechanism that has been validated by a vast accumulation of evidence, and since Darwin the emphasis in evolutionary studies has shifted to the molecular level, to genetics and the sequencing of entire animal genomes.

In years to come there can be no doubt that the life sciences in general, including studies of animals, will continue to expand. This will create career opportunities for researchers and teachers. The discoveries they make and the knowledge they transmit to future generations will be of immense benefit to the world's animals and to the world's peoples.

GLOSSARY

adaptation   the physiological, behavioral, or inherited characteristics that fit an animal for life in a particular environment.

ailurophile   a person who loves cats.

ailurophobe   a person who hates cats.

allele   one of two or more forms of a **gene** that occur in the same location on homologous **chromosomes.**

amino acid   the basic building block of proteins, consisting of an acid carboxyl group (COOH) linked to a basic amino group (NH$_2$). There are 20 commonly occurring amino acids and a few less common ones.

analogy   a dissimilarity in the structure of organs or limbs that perform similar functions in two species, the dissimilarity suggesting the organs or limbs were not inherited from a shared ancestor.

animalcule   according to the theory of **preformation,** a tiny animal present in either a spermatozoon or ovum that will grow into the juvenile form of the animal after fertilization.

anthropomorphism   the attribution of human motivation to nonhuman animals, based on the supposition that a nonhuman behaving in a way reminiscent of human behavior must be doing so for similar reasons deriving from similar thought processes. In most cases the basic supposition cannot be substantiated.

bestiary   a medieval "book of beasts," containing brief descriptions of animals, often illustrated, and drawing a moral or religious lesson from many of the descriptions.

binomial nomenclature   the standard scientific method for naming biological organisms, using one name for the genus and another for the species within that genus. The names are in Latin and are conventionally printed in italic, with an initial capital letter for the generic name and an initial lower-case letter for the specific name. For example, the domestic dog is *Canis familiaris.*

blending inheritance   the theory that offspring inherit the traits of both parents, and these merge to form an intermediate condition.

capillaries   minute blood vessels with walls only one cell layer thick from which oxygen and nutrients diffuse into tissues and into which waste material enters for later excretion. Capillaries link the arterial system carrying blood from the heart to the venal system carrying blood toward the heart.

carpals the bones linking the **radius** and **ulna** with the **metacarpals** in the vertebrate forelimb; in humans the wrist bones.

chromosome a thread of protein carrying **genes** that is found in the nucleus of all cells.

cladistics the technique of producing a diagram showing the evolutionary relationships among species.

classical conditioning (Pavlovian conditioning) a type of learning discovered by I. P. Pavlov in which an animal comes to associate an **unconditional stimulus** with a **conditioned stimulus.**

Clever Hans effect apparently intelligent responses made by an animal that in fact are influenced by unconscious signals from the questioner.

cloaca a single orifice into which the rectum and urinogenital system open; it is found in most vertebrates, but not in mammals except for the Monotremata (platypus and echidnas).

codon a sequence of three nucleic-acid bases that codes for a particular **amino acid.**

comparative anatomy the study of the anatomical differences and similarities between different species.

conditioned stimulus a neutral stimulus such as a sound that an animal learns to associate with an **unconditional stimulus.**

dance language the ritualized behavior by which a worker bee on its return to the hive informs other workers of the direction and distance to a source of food that the returning bee has located; see **round dance** and **waggle dance.**

DNA deoxyribonucleic acid, the hereditary material that carries the genetic code.

embryo a young animal during the time it is developing inside an egg or in its mother's body.

embryology the scientific study of the development of animal embryos.

entomologist a biologist who specializes in the study of insects (**entomology**).

entomology see **entomologist.**

enzyme a protein that causes or accelerates a chemical reaction in a living cell, without being altered itself.

epigenesis the theory that **embryos** develop through changes in the substances from which they are made.

ethology the study of animal behavior.

eugenics from the Greek words *eu* meaning "good" or "well" and *genes* meaning "born," a movement in the late 19th and early 20th centuries to improve the quality of human populations through selective breeding and preventing those with what were thought to be undesirable traits from having children.

eukaryotic cell a cell with a distinct nucleus enclosed by a double membrane and with other discrete cellular structures.

event recorder a portable electronic device with a keyboard that a researcher uses to type in coded observations of behavior.

flora all of the plant species found in a specified area.

fundatrix a wingless first-generation aphid **nymph.**

gastrovascular cavity in some invertebrate animals including hydra, a space containing a complex network of channels that circulate nutrients and blood.

gemmules (pangenes, plastitudes) tiny particles produced by every cell in the body that contribute hereditary information to the eggs and sperm cells in the theory of **pangenesis** proposed by Darwin.

gene the basic unit of heredity.

genetics the scientific study of biological inheritance.

genome the complete set of genetic code that an offspring inherits from its parents.

genotype the genetic constitution of an organism.

germ plasm hereditary material that passes to the eggs and sperm cells and from there to offspring, where it gives rise to the cells of the offspring's body.

hectocotylus a specialized tentacle used by male cephalopods to store sperm and to transfer it to the female's body during mating.

herbarium a book of dried plant specimens.

heterozygous possessing different copies of an **allele** (e.g., *Aa*).

histology the scientific study of biological tissues.

homology the structural similarity between organs or limbs in different species that suggests they were inherited from a shared ancestor; **chromosomes** also occur in homologous pairs that separate during cell division.

homozygous possessing identical copies of an **allele** (e.g., *AA*).

homunculus   in the theory of **preformation,** a tiny person contained inside either a spermatozoon or ovum that will grow into a baby following fertilization.

humerus   upper bone of the vertebrate forelimb.

ichthyology   the scientific study of fishes.

imprinting   a type of rapid learning that occurs early in life, in which a young animal learns to direct its social attention to a particular object, usually its mother.

larva   the immature form of an animal; it is mobile and capable of feeding itself.

mandible   in arthropods, the pair of mouth-parts used to seize prey; in birds, the lower jaw and bill or both sections of the bill (upper and lower mandible); in other vertebrates, the lower jaw.

meiosis (reduction cell division)   cell division undergone by cells involved in reproduction, in which the cell nucleus divides twice to produce four daughter cells each containing half of the original number of **chromosomes** and genetic material is exchanged between **homologous** chromosomes.

menagerie   a collection of wild or exotic animals that is held in captivity for entertainment or as a demonstration of the wealth and power of their owner.

metacarpals   the bones between the **carpals** and **phalanges** in the vertebrate forelimb; in humans, the bones of the hand.

metaphysics   the branch of philosophy concerned with ideas of being, knowing, and the mind.

mitosis   normal cell division, in which a cell divides to produce two identical daughter cells.

molecular biology   the study of the structure and functions of large molecules involved in biological processes, and especially DNA and RNAs.

Morgan's canon   in no case may we interpret an action as the outcome of a higher mental faculty, if it can be interpreted as the exercise of one that stands lower in the psychological scale.

mutation   a change in the structure of a gene or chromosome.

nymph   the **larva** of an insect in which the wings develop gradually, and that develops into the adult form with no abrupt change.

Occam's razor (principle of parsimony)   one should not increase, beyond what is necessary, the number of entities required to explain anything.

orthology the possession by different species of homologous (see **homology**) genes.

ovipary reproduction, found in most invertebrates and many vertebrates (e.g., birds), in which the female lays eggs and the young develop inside those eggs, outside the female's body.

ovovivipary reproduction, found in many insects, fishes, and reptiles, in which the young develop inside eggs that are retained within the female's body but separated from it by egg membranes, and emerge live.

pangene see **gemmule.**

pangenesis the mechanism of heredity proposed by Darwin, in which gemmules contribute heritable information to the eggs and sperm cells.

papyrus a type of paper made from the papyrus plant (*Cyperus papyrus*), which is a sedge that grows 5–9 feet (1.5–2.7 m) tall; the Egyptians were using it by about 3000 B.C.E.

paralogy the possession by a species of both copies of homologous (see **homology**) genes that underwent duplication in an ancestral species.

parthenogenesis reproduction without mating.

Pavlovian conditioning see **classical conditioning.**

phalanges the outermost bones of the vertebrate limb.

phenotype the physical appearance of an organism.

pheromone a chemical substance released by an animal into the environment that induces a behavioral response in another animal of the same species.

phylogeny the branching pattern of descent from a common ancestor to descendant species.

plastitude see **gemmule.**

preformation the theory that **embryos** develop through the unfolding of characteristics that were present throughout.

pre-Socratic a school of philosophy that existed before Socrates (ca. 470–399 B.C.E.); its principal members were Thales (ca. 624–546 B.C.E.), Anaximander (ca. 610–546 B.C.E.), Anaximenes (died 528 B.C.E.), Heraclitus (ca. 535–ca. 475 B.C.E.), Parmenides (ca. 520–ca. 450 B.C.E.), Empedocles (ca. 490–ca. 430 B.C.E.), Pythagoras (ca. 570–ca. 500 B.C.E.), Anaxagoras (ca. 500–428 B.C.E.), Democritus (ca. 460–380 B.C.E.), and Diogenes (ca. 412–323 B.C.E.).

principle of parsimony see **Occam's razor.**

pyroclastic flow   a volcanic eruption in which a cloud of hot ash, carrying pumice and other rocks, sweeps down the side of the volcano at great speed.

radio collar   a collar with a small radio transmitter fixed to it that is used to track animal movements.

radius   one of the two bones of the lower forelimb in vertebrates.

reduction cell division   see **meiosis.**

RNA   ribonucleic acid, a substance involved in the synthesis of **amino acids** from instructions carried in **DNA**; in some organisms it is the genetic material.

rostrum   a forward extension of the head, or in aphids a projection between the first pair of legs that is used to pierce plant tissues and suck sap.

round dance   in the **dance language** of bees, a behavior in which a bee returning to the hive performs a dance by moving in circles; this indicates to other bees that it has located a food source close to the hive; compare **waggle dance.**

sequencing   arranging **genes** in their correct order to form a **genome.**

spontaneous generation   the idea, prevalent in the Middle Ages, that living animals could emerge from putrefying matter.

taxonomy   the scientific classification of organisms.

teleology   the idea that an object is made for a specific purpose.

tropism   the directional response of an organism to a stimulus.

ulna   one of the two bones of the lower forelimb in vertebrates.

unconditional stimulus   a significant stimulus such as the sight or smell of food.

universal genetic code   the list of **amino acids** and the **codons** that code for them.

vivipary   reproduction in which the young develop inside the mother's body and are nourished by it until they are sufficiently developed to be born live.

waggle dance   in the **dance language** of bees, a behavior in which a bee returning to the hive informs other bees of a food source it has located by performing a dance with a figure-of-eight pattern and a zigzag or waggle path between the two halves of the eight. The length of the waggle path and the number of changes in direction indicate the distance to the food source. The angle between the waggle path and the vertical are the same as the angle between the Sun and the direction of the food if the dance is performed on the vertical comb and if it is performed on a horizontal surface outside the hive the direction of the waggle path points in the direction of the food.

zoo   abbreviation for **zoological garden.**

zoological garden (zoo)   a collection of animals that is open to the public, but that is maintained primarily for purposes of zoological research, conservation, and captive breeding programs.

zoology   the scientific study of animals.

FURTHER RESOURCES

BOOKS AND ARTICLES

Alcock, John. *Animal Behavior.* 8th ed. Sunderland, Mass.: Sinauer Associates, 8th ed. 2005. The most recent edition of a textbook on animal behavior that has become a classic.

Allaby, Michael, and Derek Gjertsen. *Makers of Science.* New York: Oxford University Press, 2002. A set of five volumes containing brief biographies of a total of 32 famous scientists, with outlines of their work.

Barber, Richard (translator and author of introduction). *Bestiary.* London: Folio Society, 1992. An English translation with all the original illustrations of M.S. Bodley 764, a manuscript held at the Bodleian Library, University of Oxford.

Beadle, Muriel. *The Cat: History, Biology, and Behavior.* New York: Simon and Schuster, 1977. A comprehensive description of the long relationship between cats and people.

Blunt, Wilfrid. *Linnaeus: The Compleat Naturalist.* London: Francis Lincoln, 2004. A full biography of Linnaeus.

Bowler, Peter J. "Darwin's Originality." *Science* 323 (January 9, 2009): 223–226. An essay explaining in simple language the central points of Darwin's theory.

———. *The Fontana History of the Environmental Sciences.* London: Fontana Press, 1992. A detailed history of all the environmental sciences.

Gruber, Jacob W. "Does the platypus lay eggs? The history of an event in science." *Archives of Natural History,* 18 (1) (1991): 51–123. An account of the first encounters between British naturalists and the platypus.

Huxley, Julian. *Evolution: The Modern Synthesis.* 1st ed. London: George Allen & Unwin, 1942. The book that merged Darwinian natural selection with Mendelian genetics in the first statement of modern evolutionary theory.

Jardine, Lisa. *Ingenious Pursuits: Building the Scientific Revolution.* London: Little, Brown and Company, 1999. A very readable account of the development of scientific ideas during the 15th to 18th centuries by a leading historian.

Kisling, Vernon N., Jr., ed. *Zoo and Aquarium History: Ancient Animal Collections to Zoological Gardens.* New York: CRC Press (Taylor and Francis), 2000. A comprehensive introduction to the history of animal collections of all kinds by an international team of authors.

Macdonald, David W. *The Encyclopedia of Mammals.* New York: Oxford University Press, 2006. A comprehensive description of mammals compiled by the professor of wildlife conservation at Oxford University.

Manning, Aubrey, and Marian Stamp Dawkins. *An Introduction to Animal Behaviour.* 5th ed. Cambridge: Cambridge University Press, 1998. A straightforward, clearly written textbook on animal behavior.

Mason, Peter. *Deconstructing America: Representations of the Other.* New York: Routledge, 1990. An account of the search for fantastic humans, similar to those of Eurasian mythology, during the early exploration of the Americas.

Mullan, Bob, and Garry Marvin. *Zoo Culture.* Chicago: University of Illinois Press, 1998. A study of zoos and their history.

Ogilvie, Brian W. *The Science of Describing: Natural History in Renaissance Europe.* Chicago: University of Chicago Press, 2006. A scholarly account of the development of natural history in Europe.

Payne, Ann. *Medieval Beasts.* London: The British Library, 1990. A modern bestiary compiled from original medieval manuscripts and illustrations.

Polo, Marco. *The Travels,* translated by Ronald Latham. New York: Penguin Books, 1958. The text of Marco Polo's book with an introduction by the translator.

Ridley, Mark. *Animal Behavior.* 2nd ed. Malden, Mass.: Blackwell Science, 1995. An introduction to animal behavior with many examples and descriptions of experiments.

Russell, Bertrand. *History of Western Philosophy.* London: George Allen & Unwin, 1946. A book by one of the greatest philosophers of the 20th century that explains clearly the ideas of the Greek philosophers (and of philosophers up to the early 20th century).

Skelton, Peter, ed. *Evolution: A Biological and Palaeontological Approach.* Harlow, Britain: Addison-Wesley in association with the Open University, 1993. An undergraduate-level textbook on evolution and genetics.

White, David Gordon. *Myths of the Dog-Man.* Chicago: University of Chicago Press, 1991. A scholarly account of the myths of strange, partly human beings from around the world.

WEB PAGES

Aristotle. *On the Generation of Animals.* Translated by Arthur Platt. Available online. URL: http://etext.library.adelaide.edu.au/a/aristotle/

generation/index.html. Accessed November 28, 2008. The full text of Aristotle's work.

———. *The History of Animals.* Translated by D'Arcy Wentworth Thompson. Available online. URL: http://etext.library.adelaide.edu. au/a/aristotle/history/. Accessed November 28, 2008. The full text of Aristotle's work.

Arnott, Michael, and Iain Beavan. "Bestiary" University of Aberdeen. Available online. URL: http://www.abdn.ac.uk/bestiary.hti. Accessed January 6, 2009. An account of the history and content of medieval bestiaries.

Baltzly, Dirk. "Stoicism." In *Stanford Encyclopedia of Philosophy.* Available online. URL: http://plato.stanford.edu/entries/stoicism/. Accessed November 25, 2008. A brief but comprehensive account of Stoic philosophy.

Bateson, Patrick. "William Bateson: A Biologist Ahead of His Time." *Journal of Genetics* 81, no. 2 (August 2002). Indian Academy of Sciences. Available online. URL: http://www.ias.ac.in/jgenet/Vol81No2/49.pdf. Accessed January 29, 2009. A brief biography of Bateson (Prof. Patrick Bateson is not a direct descendent, although he is distantly related to William Bateson).

Cicero, Marcus Tullius. *On the Nature of the Gods.* Translated by Francis Brooks. Available online. URL: http://oll.libertyfund.org/ ?option=com_staticxt&staticfile=show.php%3Ftitle=539#toc_list. Accessed November 26, 2008. The full English-language text of Cicero's *De Natura Deorum.*

International Commission on Zoological Nomenclature. "International Code of Zoological Nomenclature." Available online. URL: http:// www.iczn.org/iczn/index.jsp. Accessed January 23, 2009. The official code governing the scientific classification of animals.

Lendering, Jona. "Pliny the Elder." Available online. URL: http://www. livius.org/pi-pm/pliny/pliny_e.html. Accessed December 1, 2008. A biography of Pliny and description of the *Natural History.*

Lloyd-Jones, Sir Hugh. "Ancient Greek Religion." Available online. URL: http://www.aps-pub.com/proceedings/1454/405.pdf. Accessed November 21, 2008. A brief outline of the origin of Greek religion by the Regius Professor of Greek, Emeritus, Oxford University.

Maienschein, Jane. "Epigenesis and Preformationism." In *Stanford Encyclopedia of Philosophy.* Available online. URL: http://plato.stanford. edu/entries/epigenesis/. Accessed January 23, 2009. An outline of the history of the theories of epigenesis and preformation.

Mendel, Gregor. "Experiments in Plant Hybridization." Available online. URL: http://www.mendelweb.org/Mendel.html. Accessed January

28, 2009. The full text of Mendel's original paper, translated into English.

Natural History Lecture Service. "Zoological Classification and Names." Available online. URL: http://www.theanimalman.co.uk/animfacts/classification%201.html. Accessed January 23, 2009. An explanation and full list of the names used in classifying animals down to orders.

Nobel Foundation. "Ivan Pavlov." *Nobel Lectures.* Amsterdam: Elsevier Publishing Co. Available online. URL: http://nobelprize.org/nobel_prizes/medicine/laureates/1904/pavlov-bio.html. Accessed February 4, 2009. Biography of Pavlov.

———. "Konrad Lorenz." *Nobel Lectures.* Amsterdam: Elsevier Publishing Co. Available online. URL: http://nobelprize.org/nobel_prizes/medicine/laureates/1973/lorenz-autobio.html. Accessed February 4, 2009. Autobiography by Lorenz.

———. "Nikolaas Tinbergen." *Nobel Lectures.* Amsterdam: Elsevier Publishing Co. Available online. URL: http://nobelprize.org/nobel_prizes/medicine/laureates/1973/tinbergen-autobio.html. Accessed February 4, 2009. Autobiography by Tinbergen.

O'Connor, J. J., and E. F. Richardson. "Georges Louis Leclerc Comte de Buffon." Available online. URL: http://www-groups.dcs.st-and.ac.uk/~history/Biographies/Buffon.html. Accessed January 26, 2009. A brief biography of the comte de Buffon.

———. "René Descartes." Available online. URL: http://www-groups.dcs.st-and.ac.uk/~history/Biographies/Descartes.html. Accessed January 20, 2009. A brief biography of Descartes.

O'Keefe, Tim. "Epicurus (ca. 341–271 B.C.E.)." In *Internet Encyclopedia of Philosophy.* Available online. URL: http://www.iep.utm.edu/e/epicur.htm. Accessed November 25, 2008. Biography of Epicurus and explanation of his philosophy.

Wildman, Wesley. "Pre-Socratics (600–400 B.C.E.)" Available online. URL: http://people.bu.edu/wwildman/WeirdWildWeb/courses/wphil/lectures/wphil_theme01.htm. Accessed November 24, 2008. A concise explanation of who the pre-Socratic philosophers of ancient Greece were, what they believed, and what came after them.

INDEX

Note: *Italic* page numbers indicate illustrations; page numbers followed by *m* indicate maps.